NatureServe
A Network Connecting Science With Conservation

At NatureServe, we believe conservation efforts should be guided by strong scientific data. We are the leader in providing scientific information, tools, and expertise to help guide effective conservation action in the Western Hemisphere. Our efforts have established NatureServe as the go-to source for information about rare and endangered species and threatened ecosystems.

Biodiversity Science

Biodiversity science is the foundation of NatureServe. Our mission is to collect, manage, analyze, and communicate high-quality and reliable information about species and ecosystems.

Costa's Hummingbird (*Calypte costae*)

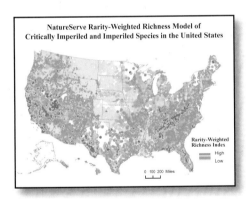

Conservation Tools & Services

We develop standards and methods for collecting scientific data. Our customized data requests and nine interactive online data tools help you take action where it's needed most. NatureServe's experienced staff will help you plan your projects and meet your objectives by interpreting data and guiding action.

NatureServe Network

Our network consists of more than 80 member programs and more than 1000 conservation professionals working from Alaska to Peru.
The NatureServe network collects, analyzes, and distributes detailed scientific data on more than 70,000 species and 7000 ecosystems.

Learn more about NatureServe by visiting

www.natureserve.org

Winner of the MacArthur Award for Creative and Effective Institutions

FOR NATURE
FOR NOW
FOREVER

PHOTO BY LIZ SAUNDERS.

Give a gift today and help protect
Canada's most vulnerable habitats and
the plants and animals they sustain.

DONATE NOW or learn more: 1-800-465-8005 🍁 natureconservancy.ca

Being a Bird in North America,

NORTH OF MEXICO

Volume 1:
Waterfowl to Shorebirds

ROBERT ALVO

RECOMMENDED CITATION:
Alvo, R. 2015. *Being a Bird in North America, North of Mexico (BABINA)*, Volume 1: *Waterfowl to Shorebirds*. Ottawa: Robert Alvo.

Copyright © 2015 Robert Alvo
Published by Robert Alvo

RESEARCH AND TEXT:
Robert Alvo

IDEAS FOR CARTOONS:
Robert Alvo, Robert L. Manson

CARTOON REVIEW:
Robert L. Manson, Bob Bracken
(deceased), Josée Nesdoly

MAPS:
Josée Nesdoly (quality control by
Robert L. Manson and Robert Alvo)

SPANISH BIRD NAMES FOR MEXICO:
Héctor Gómez de Silva

**CONSERVATION STATUS N-RANKS
    FOR MEXICO:**
Héctor Gómez de Silva, Mónica
Pérez Villafaña, and Robert Alvo

CONTENT REVIEW:
Raleigh J. Robertson, Robert L. Manson,
Jean-Pierre L. Savard

PROOF-READING:
Fred Helleiner, Leslie Alvo,
Robert L. Manson, David Seburn

FRONT COVER ART:
Christina Lewis

LAYOUT DESIGN:
Robert Alvo, Robert L. Manson,
Josée Nesdoly, Suzanne Burkill

DIGITAL LAYOUT AND FORMATTING:
Suzanne Burkill

**COLOR CORRECTION AND
    DIGITAL IMAGE MANIPULATION:**
Elizabeth Payne, Suzanne Burkill

INDEX:
Gillian Watts

WEBSITE:
Robert Alvo

Printed and Bound in Canada by Friesens.

*Library and Archives Canada
Cataloguing in Publication*

Alvo, Robert, 1959–
Being a Bird in North America,
    north of Mexico / Robert Alvo.
Includes bibliographical references
and index.
Contents: v. 1. Waterfowl to shorebirds.
ISBN 978-0-9877733-0-2 (v. 1)

1. Birds – Canada – Identification.
2. Birds – United States – Identification.
3. Birds – Ecology – Canada.
4. Birds – Ecology – United States.
5. Birds – Canada – Geographical
    distribution.
6. Birds – United States – Geographical
    distribution.
I. Title.

QL681.A48 2013   598.097
    C2012-906638-9

This book is dedicated to all forms of life,
from lowly bacteria to primates,
and especially to extraterrestrial forms,
who are all wondering
what on Earth humans are doing to this planet.

RA

Virtual Moon

It's only a virtual moon
And an electronic sea,
But it wouldn't be make-believe
If you would talk to me.

It's a bits and bytes parade.
It's a melody seen
On a plastic screen
And in a hard-drive played.

It's an electronic moon
In a cybernetic sea
And I wish it was make-believe
And not reality.

Reprinted by permission of Mr. Murray Citron

Murray Citron's *Virtual Moon* is the current reality. However, many people still prefer the paper book, especially for such an image-intense work as this one. My dream all along the process has been to create a long-lasting paper book, and this has not changed. After publication, I may consider producing an electronic version if the demand exists. At the time of writing, I understand that the proportion of the book market that electronic books command in North America has peaked and is in decline.

RA

Acknowledgments

ADVERTISEMENTS: Kei Sochi (American Birding Association), Elaine Secord (Bird Studies Canada), Elizabeth Sbaglia (Nature Conservancy of Canada), Brian Cardillo (NatureServe).

BOOK-TITLE ADVICE: Steve Wendt.

CITATIONS, QUOTATIONS, AND REFERENCES ADVICE: Alan Poole, Michael Berrill, Mike Runtz.

CONSERVATION STATUS OF SPECIES ADVICE: Syd Cannings, Myke Chutter, Lea Gelling, Tracey Gotthardt, Colin Jones, Bill Pranty.

DATA STORAGE: Suzanne Burkill, Robert L. Manson, Sylvie Marchand, Venkatesh Sosle.

EVENTS TO PRESENT/SELL THE BOOK: Lynn Clark.

FINANCIAL, TAX: Joe Coneybeare.

GENERAL ADVICE: Frances Cook, Stuart Houston, Bob Nero, Stewart Peck.

INTERNATIONAL ISSUES ADVICE: Jim Lawrence, Nick Prentice (both with Birdlife International).

LAYOUT ADVICE: Elizabeth Payne, Barbara Robinson.

LEGAL ADVICE: Byron Pascoe.

LIBRARY SERVICES: Staff at the Canadian Wildlife Service national library and the University of Ottawa library.

LIFE-BIRD FINDING: Oliver Ashford, San Blas, MX; Grant Beauprez, NM; Bob Behrstock, se. AZ; Bob Bledsoe, ne. CO; Dick and Russ Cannings, Okanagan Valley, BC; Bonnie Chartier, Churchill, MB; Forrest Davis, AK; Jose Inocencio Bañuelos Delgado (Chencho), San Blas, MX; Barbara MacKinnon de Montes, Yucatan, MX; Bruce Di Labio, Ottawa, ON; Gjon Hazard, San Diego, CA (pelagic tour); Alfredo Herrera, Puerto Vallarta, MX; Bernie Ladouceur, Ottawa, ON; Braulio Mállaga, Ruiz Cortines, MX; Thor Manson, Okanagan Valley, BC; Bryan J. Smith, c. AZ; Bob Stewart, se. AZ; Glenn Walbeck, CO; Bruce Whittington, s. Vancouver Island, BC; Joe Woodley, se. AZ.

MAP PERMISSIONS: Liz Gordon, Josep del Hoyo, Rob Riordan.

NATURESERVE ASSISTANCE: Nicole Capuano, Erin Chen, Pat Comer, Kyle Copas, Leslie Honey, Michael Jewell, Colin Jones, Don Kent, Mary Klein, Pat Leighty, Larry Master, Jason McNees, Margaret Ormes, Ravi Shankar, Bruce Young.

PHOTO ASSISTANCE: Dawna Jones, Edgar T. Jones (deceased) and Jeanne Jones (deceased), Dave Moore, Mats O. G. Eriksson, Robert L. Manson, Gilles Seutin.

PRINTING ADVICE (FRIESENS): Don Feaver, Marg McLeod, Brad Schmidt.

QUALITY CONTROL OF SPECIES DATA: Véronique Alvo.

REVIEW OF SOME SPECIES ACCOUNTS: Bob Bracken (deceased), Heidi Cline, Tom Cooper, Héctor Gómez de Silva, John Eadie, Sue Haig, Stuart Houston, Christina Lewis, Patrick Magee, Mark Mallory, Frank Phelan, Michel Robert, Cliff Shackelford, Rodger Titman.

SELF-PUBLISHING, MARKETING, SELLING ADVICE: Jim Duncan, Dawna Jones, Dennis Paulson, Bob Stewart, Paul Valliant.

SOFTWARE AND HARDWARE ADVICE: Apple Store, Ottawa. Howard Ko, Laurier Computer.

SPANISH TRANSLATIONS: Kate Alvo.

SPECIES VERIFICATION IN PHOTOS: Michel Gosselin.

SYSTEMATICS, TAXONOMY AND NOMENCLATURE ADVICE AND DATA: R. Terry Chesser, Michel Gosselin.

My sincere apologies to any folks I may have failed to mention.

Contents

Introduction

Who This Book is For

This book is intended for a wide audience, which includes birders, anyone with a passing interest in birds, nature lovers, anyone interested in conservation, or simply people who enjoy humor. Each species page is a lesson in itself and can be used for classes in environmental studies from perhaps Grade 9 and upwards, although younger children will still enjoy the cartoons.

Nature and Life

I've always been fascinated by nature, especially living nature, or biodiversity, and have come to understand the *elements of biodiversity* as species and ecosystem types, particularly for the purposes of biodiversity conservation.

In a fast-moving world that depends on biodiversity, we must make priorities in terms of which places on the landscape (or in the ocean) to protect. It can be very useful to work at different jurisdictional levels (e.g., a state), because they are the ones at which decisions are made about the landscape. In order to make lists of species and ecosystems in the chosen jurisdiction, taxonomy based on a logical classification system is required, and these classifications are usually hierarchical.

Most people understand the concepts of species, species lists, and ranking species for conservation purposes. And most people know that different kinds of living nature, or ecosystem types, are groups of the same dominant species interacting with their environment. However, defining ecosystem types and recognizing that they also fit into a hierarchically arranged taxonomy are not as evident. NatureServe and the broader scientific community have created a hierarchically organized ecosystem classification for North America. The terrestrial portion describes forests, woodlands, shrublands, grasslands, wetlands, and sparsely vegetated lands, whereas the freshwater portion describes rivers, streams, and lakes. The subterranean portion describes the different kinds of caves, and a marine portion also exists. Thus broad-scale formations (e.g., forest) have been organized hierarchically, from generalized categories such as "cool temperate forest" down to more narrowly defined plant communities, like a "Northern Red Oak - Yellow Birch / Cinnamon Fern Forest." (www.natureserve.org/conservation-tools/standards-methods/ecosystem-classification) (http://usnvc.org) (http://cnvc-cnvc.ca/).

The existence of this ecosystem taxonomy makes it possible to name ecosystem types, as is done with species, and also rank them for conservation concern (see "NatureServe Conservation Status Ranks" below). Protecting some proportion of each ecosystem type in a network of conservation areas protects much of the biodiversity it contains, including species that are too little known and too numerous to work on individually (e.g., most invertebrate groups, most non-vascular plant groups, fungi, lichens, algae, single-celled organisms). This coarse-filter approach to protecting biodiversity conserves the space and ecological processes needed by species as the environment changes and as everything evolves.

However, because some species become vulnerable to specific human influences, the coarse-filter approach must be complemented by conservation actions for individual species. This is the so-called "coarse-filter/fine-filter approach" to protecting biodiversity. The most important places in the jurisdiction to protect (or to recognize as already being protected) can then be prioritized based on how representative they are of the natural communities they contain, and also the number of high-ranking elements of biodiversity, the rank of each element, and the condition of each element.

Why Birds?

So, why focus on birds? They aren't the most important group ecologically, the most numerous group in terms of species or numbers of individuals, or even necessarily the most interesting group. Birds are simply one of the most popular groups of life to humans. This is evident in the amount of scientific and popular literature available (current and historic), the number of conservation dollars directed toward birds, the amount of attention paid to them in environmental impact assessments, and the number of organizations devoted to birds. Birds are also the group I know best, so it makes sense to start with them in developing a book model that can be used for the other elements of biodiversity.

What this Book is Not

A question I often get asked is, "Why another book on birds?" To answer this, it helps to describe what it *is not*. It's not a field guide, of which I own many. Nor is it a bird-finding guide, such as the many that I own that have given me specific directions to "birdy" sites. It's not a description or summary of what is known about each species, such as Arthur Cleveland Bent's 21-volume *Life Histories of North American Birds* series (e.g., Bent 1919), the *Handbook of the Birds of the World* accounts (del Hoyo et al. 2015), or even the *Birds of North America Online* (http://bna.birds.cornell.edu/bna) accounts. (All three of these sources have been critical references for the current book.) It's not a poetic or romantic book about birds, nor is it a coffee-table book. It's not a state or provincial atlas.

What this Book Is

The various types of books on birds mentioned above make up a good portion of the popular literature on birds. So if this book is none of these, what is it? Well, it has aspects of all of them, and it is the first in a series of books whose main theme is conservation. A basic premise is that each element (or species, in the case of birds) can be used to tell a story, where each story can reveal the tricks used by birds to survive, describe real problems faced by birds, analyze conservation issues, or illustrate ecological characteristics or something else that applies to— or is unique about—the species. This book also provides a way of remembering and distinguishing each species. For most people, pictures help to understand ideas and make it easier to store them away in memory. And in some ways cartoons are the best kind of picture for this purpose, because they elaborate, exaggerate, and—if done well—make you laugh.

Geographical Scope

The area covered by this book is the area encompassed by the American Birding Association (Pranty et al. 2008): "the 49 continental United States, Canada, the French islands of St-Pierre and Miquelon, and adjacent waters to a distance of 200 nautical miles from land or half the distance to a neighboring country (i.e., the Bahamas, Cuba, Greenland, Mexico, or Russia), whichever is less. Excluded by these boundaries are Bermuda, the West Indies, Hawaii, and Greenland." Also excluded are US islands outside of the 200 nautical mile limit (e.g., Mariana Islands, Wake Island, Johnston Island). Thus, this book covers North America, north of Mexico, as stated in the title. For brevity, this is shortened to "North America" in the text. Statements such as "*Our* seven grebe species" mean "the seven grebe species breeding regularly in North America", i.e., the ones being treated in the main section of this book.

It seems a shame not to include Mexico in the geographic scope, given it is such a large neighbor of the US, but doing this would have made the number of species unmanageable. Rather I have coyly made it the only country named in the title of the book, provided Spanish bird names for use in Mexico, and presented conservation status ranks for Mexico. It is shaded lightly in the map on the book's inside front cover to reflect its partial inclusion.

Which Species to Include?

Almost 1000 bird species are known in the wild in North America. The selection of species to cover is based on their conservation status in North America, such that some are treated in the main section of the book on their own species pages, while the rest are simply given a line in the Appendix. The first group includes all native species that breed, or used to breed (i.e., species that became extinct or extirpated since roughly the time of European colonization of North America), with some regularity in North America. The second group includes all other species that have been documented to occur at least once, but with no evidence of regular breeding, namely: regularly occurring nonbreeders, and birds of accidental or casual occurrence. Also included in the second group are species introduced by humans to North

America that have become established (i.e., breed regularly). In other words, the main section of the book treats the species that can be, and should be, protected, whereas the Appendix lists those that either cannot be, or should not be, protected in North America.

Not included in either section of the book are species known in North America only from records of hybrids, feral species, escaped individuals, introduced species that have not become established, and other birds about which there is doubt regarding their identity or origin. Decisions on the section of the book in which to include each species are based on distributional information found in American Ornithologists' Union (AOU) (1998), as updated to 2014 with supplements to the journal *Auk* (including the 55th supplement (Chesser et al. 2014)). Grouping the species using this method yields approximately 670 species for the main section of the book and approximately 320 species for the Appendix (the exact number will likely change before Volumes 2 and 3 are published).

Species Order

The bulk of this book consists of species accounts (one species per page) arranged in the taxonomic order defined by the AOU. As above, the classification (including species order) and nomenclature used in the current book is up-to-date to 2014 (the 55th supplement). The only exception is the Clapper Rail (see account, p.155, for explanation). The 2014 AOU check-list is also available on-line at: www.aou.org. Given the one-species-per-page format and the number of species to treat in the main section, I decided to publish the book in three volumes, each with at least 200 species in its main section. Volume 1 treats 206 species.

Species Pages

All the species pages contain the same elements: names section, cartoon, global distribution map, NatureServe conservation status, photo(s), and a text account.

Names Section

The species' English name makes up the page's header. The scientific name, French name ("*Français*"), Spanish name for Mexico ("*México*") (Spanish bird names vary from country to country and even within one country), taxonomic order, and family are presented in the names section. With the exception of the Spanish name for Mexico, for which Berlanga et al. (in prep.) is followed, these names follow the AOU. (Spanish species names for Mexico that do not appear in that book are provided courtesy of Dr. Héctor Gómez de Silva.) In the few cases for which French names are not given by the AOU, Devillers et al. (1993) prevails.

Cartoons

Each cartoon was created specifically for this book. Fifteen artists, each with their own distinctive style, contributed, and the media that they used also vary widely.

Maps

The purpose of the global range maps is to give the reader an idea of where the species breeds, where it winters, and where it occurs as a year-round resident. We have not included the migration range. Apparently only one other paper publication provides global range maps for all the North American birds: the series of 17 books edited by Dr. Josep del Hoyo and entitled, "Handbook of the Birds of the World" (HBW), published from 1992 to 2013. The entire series is now available online under the title "HBW Alive", along with additions (e.g., videos, photos, recordings of vocalizations, updated checklist), at: (www.hbw.com). NatureServe has published Western Hemisphere range maps of North American birds on its website (www.natureserve.org) (Ridgely et al. 2005). With permission from both Dr. del Hoyo and NatureServe, global range maps were drawn using NatureServe's maps for the New World and HBW maps for the Old World. In some cases these data were adapted to accommodate recently re-defined taxa (e.g., from species splits).

The colors used in the maps are as follows:

■ Breeding | ■ Nonbreeding | ■ Year-round resident | ■ Introduced
Staggered lines running through the range indicate areas of extirpation or extinction. For

introduced species, in any discrepancy between the map and the text, the text prevails. Introduced ranges are not always shown.

These small maps cover much ground. If you want to know your chances of seeing a given species in a region at a certain time of year, you can consult various published and online sources. Atlases usually give precision to 10 by 10 kilometers. Bird-finding guides give detailed directions for travel to specific sites for finding certain birds, and if you're interested in knowing when a particular species was last found in a region or at a specific site, http://ebird.org is indispensable, as are local hotlines and listserves. Ask local birders.

NatureServe Conservation Status Ranks

NatureServe designates conservation status ranks for species and ecosystem types to help set priorities for conservation. These ranks are an indication of the risk of extinction or extirpation, and include:

1) Critically Imperiled;

2) Imperiled;

3) Vulnerable;

4) Apparently Secure; and,

5) Secure.

Ranks are determined on three distinct jurisdictional scales: global ("G-Ranks"), national ("N-Ranks"), and subnational (i.e., state or province) ("S-Ranks"). Factors considered in ranking species include: rarity, distribution, threats, and population trends (www.natureserve.org).

Each species page presents G-Ranks (e.g., G2 for Imperiled on the Global scale) along with N-Ranks for Canada, the US, and Mexico (e.g., N2) (NatureServe 2014a). (Only G-Ranks are given in the Appendix.) For the sake of simplicity, rounded ranks are presented. For example, the global rank G3G4, which allows for uncertainty between G3 and G4, is rounded up to G3. In addition, when an N-Rank reflects both a breeding status and a nonbreeding one, as separated by a comma (e.g., N1B,N3N), the N-Rank given in this book applies only to the breeding population (e.g., N1). If only a nonbreeding population exists, the rank given here applies to it.

The numeric ranking system does not accurately reflect some species' status, and they are given a letter instead, as follows, for the global level:

GX —*Presumed Extinct* (from the Earth). Not located despite intensive searches and virtually no likelihood of rediscovery.

GH —*Possibly Extinct*. Missing; known only from historical occurrences but some hope of rediscovery remains.

GU —*Unrankable*. Not able to be ranked, due to lack of information or to substantially conflicting information about status or trends.

GNR—*Unranked*. Conservation status not yet assessed.

For the national level, letter ranks are:

NX —*Presumed Extirpated* from the nation as a breeder, but possibly occurs there as a nonbreeder.

NH —*Possibly Extirpated*. Known as a breeder only from historical occurrences, but some hope remains of rediscovery as a breeder.

NU —*Unrankable*. Not able to be ranked, due to lack of information or to substantially conflicting information about status or trends.

NNA—*Not Applicable*. A conservation status rank is not applicable because the species is not a suitable target for conservation activities. Includes species that occur as long-distance migrants or nonbreeders, others that occur casually (i.e., not every year), and accidentals (only one or a few confirmed reports). Also included are species for which no confirmed reports exist, introduced species, and hybrids. Note that species that are known only from escaped individuals that have not become established are not included as exotics. They are not included in the book.

If the only ranking criteria used at the three jurisdictional levels (G, N, S) were rarity and distribution, a higher numeric rank would not be possible at a lower jurisdictional level (e.g., G1N2). However, because threat levels and population trends can differ at different jurisdictional levels, such ranks (e.g., G4N5) do sometimes occur for wide-ranging species. In this example, the species may be declining enough at the global level to warrant a G4 rank, but the portion of the population in the nation is stable and large enough to merit an N5 rank. Higher ranks at lower jurisdictional levels can also occur temporarily when ranking at different jurisdictional levels is not simultaneous. For a good example of the use of NatureServe conservation status ranks, see the Gunnison Sage-Grouse account (p. 67).

N-Ranks for Mexico were determined in 2015 by Héctor Gómez de Silva, Mónica Pérez Villafaña, and me (RA) using the most recent NatureServe conservation status ranking methodology, i.e., the "rank calculator" (www.natureserve.org).

N.B. In this book, NatureServe conservation status ranks are usually expressed using the spelled-out forms in title caps (e.g., Apparently Secure).

Text

It is not the objective of this book to summarize all that is known about each species—the text does not have a number of headings such as identification, habitat, and behavior. Rather, the approach is to focus on a small number of topics that apply to the species. The range of possible topics is diverse, and the intent is to minimize repetition of topics between species. You can think of it as "the best of each species".

Photos

Photo captions are kept to a minimum, not only to save space, but also to avoid repetition (e.g., "breeding male"). Another reason was to avoid the problem of difficulty in separating young birds from females. The photos always show the species in question unless otherwise stated in a caption. Each photo can be seen as an identification quiz. On pages with more than one photo, the priority from left to right and top to bottom is as follows: breeding pair, breeding individual, male, female, nonbreeding, sub-adult (oldest to youngest), nest, nest site, different taxon (e.g., race). The inclusion in this book of certain photos was facilitated by the fact that the photographers were engaged in scientific research.

Style and Rules

The Chicago Manual of Style, 16th edition (University of Chicago 2010) is used as the main guide on issues of style. However, it often allows the author to determine how to treat certain situations rather than having fixed rules for every situation. Spelling conforms mostly to Merriam-Webster (2015), although some differences exist between it and "Chicago". The intent was to maintain consistency throughout the book in terms of style rules and spelling rules.

The metric system is used in this book. Based on multiples of ten, it is simpler than the English system, and on a global basis it is in much wider use. For the purposes of simplicity in understanding the general idea while reading this book, the reader unfamiliar with metric can use the following approximations:

1 meter (m) ≈ 1 yard	1 kilometer (km) ≈ 0.6 mile
1 hectare (ha) ≈ 2.5 acres	1 liter (l) ≈ 1 quart

A change (e.g., increase) of a Celsius (°C) degree is equivalent to a change (e.g., increase) in the same direction of two degrees Fahrenheit.

In the narratives, scientific names are given after a species' English name only for species (bird or other) not treated in the book, either in the main part or in the Appendix. English species names are capitalized throughout, as they are proper nouns. This approach helps avoid confusion between different possible meanings for terms such as: "yellow warbler". Was it a warbler with considerable yellow in its plumage (perhaps a Magnolia Warbler), or was it the species we call "Yellow Warbler" (*Setophaga petechia*)? This rule is applied in this book for all life forms.

List of Species*

*Species ordering is based on the most recent scientific interpretation of shared evolutionary descent and follows the American Ornithologists' Union – Chesser et al. (2014)

Black-bellied Whistling-Duck

Scientific: *Dendrocygna autumnalis*
Français: Dendrocygne à ventre noir
México: Pijije Alas Blancas
Order: Anseriformes
Family: Anatidae

Rare in the duck world is the drake who helps his mate incubate eggs and raise young. Yet this is the way in whistling-ducks, of which we have two species. Black-bellied Whistling-Ducks pair for life. Both sexes choose the nest site, approaching the cavity or nest box from above and extending the neck downward to examine it. Atypically for waterfowl, this species doesn't line the nest with down.

Incubation begins when the hen remains on the nest overnight after laying an egg, usually two to three days before laying the final egg. Embryos may start to develop before incubation due to high air temperatures. The female doesn't actually give the eggs to the male as suggested in the cartoon, but the fact that he incubates at all makes him worthy of "super-duck" status nonetheless. He even aids her in brooding the ducklings during their first 12 days of life, first at the nest, then elsewhere. After fledging, the young remain with the parents for up to four months, or possibly even longer (Dale and Thompson 2001).

Juan Bahamon

Brood parasitism, in which more than one female lays eggs in the same nest, occurs often in this species. However, its frequency is difficult to determine because females don't sign their eggs, but researchers can infer brood parasitism from the addition of more than the usual one egg per day or from a clutch size of more than 14 eggs (even 100!). Also, egg color, size or shape can differ consistently between females.

The Black-bellied Whistling-Duck's breeding range has been expanding greatly in the US from Texas and from Louisiana introduction sites (Bergen 1999) eastward to Florida (Woolfenden and Robertson 2006) and South Carolina (Keyes 2010), as well as northward to Oklahoma (Arterburn 2004).

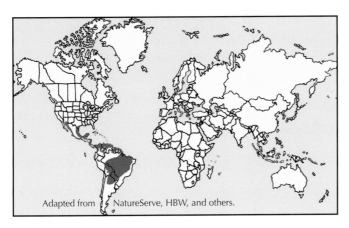

Adapted from NatureServe, HBW, and others.

NatureServe
Conservation Status

Global: .... Secure
Canada: .... Not Applicable
US: .... Secure
Mexico: .... Secure

www.natureserve.org

Fulvous Whistling-Duck

Scientific: *Dendrocygna bicolor*
Français: Dendrocygne fauve
México: Pijije Canelo
Order: Anseriformes
Family: Anatidae

Don't laugh at the idea of rice fields as bird habitat. Waterfowl, wading birds, shorebirds, cranes, rails, and even some landbirds use them. In fact, rice fields are sometimes a species' only habitat for feeding and/or nesting in some parts of the world. Rice fields represent 15% of the world's wetlands. Waste rice, spilled during harvest or not collected, is the most important food source. Other food sources include grass, sedge, and forb seeds, vertebrates, and invertebrates. Seasonal flooding and tilling create many habitats annually, from mud flats to water 30 cm deep. Bird use of rice fields is dependent on size, depth, hydrology, habitat connectivity, distance to natural wetlands, and distance to unsuitable habitat.

The Fulvous Whistling-Duck loves rice fields. It has five populations in four continents. This, and its lack of subspecies or geographic variation could make it an excellent subject for comparing rice field issues geographically. It expanded into the southern US in the 19th century and may be the least studied of the common North American waterfowl. Most studies here have focused on it as a pest that feeds mostly on seeds, often at night to avoid retaliation from farmers.

Edgar T. Jones

Birds in rice fields trample crops, graze on young plants, eat rice, and damage levees, but also improve crop yields by eating weed seeds, controlling other pests, and improving nutrient cycling by their actions that stir the water and the substrate. They also help farmers via hunting and ecotourism, and by allowing them to grow wildlife-friendly pesticide-free rice, which commands a higher price and is healthier for birds and humans. The 1,700 year-old Chinese rice-fish approach of growing fish in rice fields is recognized by the United Nations as a Globally Important Ingenious Agricultural Heritage System. A North American analog is crayfish-growing in Louisiana rice fields (Hohman and Lee 2001; Elphick et al. 2010).

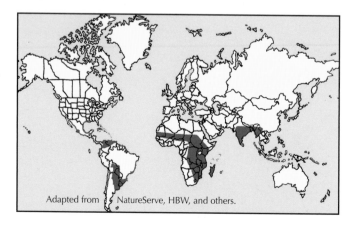

Adapted from NatureServe, HBW, and others.

NatureServe
Conservation Status

Global: .... Secure
Canada: .... Not Applicable
US: .... Apparently Secure
Mexico: .... Secure

www.natureserve.org

Greater White-fronted Goose

Scientific: *Anser albifrons*
Français: Oie rieuse
México: Ganso Careto Mayor
Order: Anseriformes
Family: Anatidae

"The peculiar laughing cry of this bird has given it the name of 'laughing goose'. Its cries are said to be loud and harsh, sounding like the syllable *wah* rapidly repeated; the note is easily imitated by striking the mouth with the hand while rapidly uttering the above sound" (Bent 1925, 193). The French name *Oie rieuse* also means "laughing goose". Roosting flocks may play by making reverse somersaults in water, wing-flapping, chasing each other in circles, and diving (Ely and Dzubin 1994).

The species name *albifrons* refers to the distinctive white band at the base of the bill, as does the word *careto* in the Mexican name, but the black blotches on the underparts, the pink bill, and orange feet are also good field marks. The similar-looking domesticated descendant of the Graylag Goose (*A. anser*), usually doesn't have the white face band.

The Greater White-fronted Goose is the only regularly breeding New World member of the eight species in the genus *Anser*, which includes: Swan Goose (*A. cygnoides*), Taiga Bean-Goose (*A. fabilis*), Tundra Bean-Goose (*A. serrirostris*), Pink-footed Goose (*A. brachyrhynchus*),

Netta Smith

Lesser White-fronted Goose (*A. erythropus*), Graylag Goose, and Bar-headed Goose (*A. indicus*) (Clements et al. 2012). Knowing how to recognize our representative gives the North American birder visiting the Old World or Alaska a solid basis to identify these other so-called "gray geese".

In autumn 1990, 90% of the continent's population (460,000 birds) graced The Galloway and Miry Bay Important Bird Area (IBA) in southwestern Saskatchewan (Smith 1996). That IBA is the most significant prairie staging area for this goose, which lucky easterners may see by scanning flocks of Canada Geese.

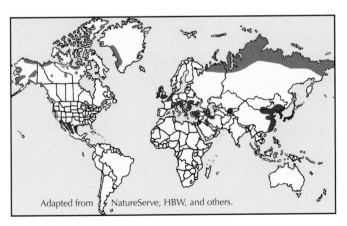

Adapted from NatureServe, HBW, and others.

NatureServe
Conservation Status

Global: .... Secure
Canada: .... Secure
US: .... Secure
Mexico: .... Not Applicable

www.natureserve.org

Emperor Goose

Scientific: *Chen canagica*
Français: Oie empereur
México: Ganso Emperador
Order: Anseriformes
Family: Anatidae

This Alaskan specialty bird has a small global range at all times of year, breeding in western Alaska and eastern Siberia and wintering mostly on the coasts of the Aleutian Islands and the Alaska Peninsula next to ice-free waters. It roosts and feeds near the water's edge, and nests in coastal salt marshes. The adult's white head and hind neck are often stained orange-red from iron in the sediment. The white eggs also are often stained, presumably from the same cause.

The global population is Vulnerable. An oil spill in its small range would be disastrous because it breeds within 15 km of the coast in flat salt marsh habitats, and it feeds and roosts in the intertidal zone. Indeed, oil-spotted birds have been seen in the Aleutian Islands, likely having come in contact with waste oil dumped from ships. The global spring population dropped from 139,000 birds in 1964 to 101,000 in 1982, then to 42,000 in 1986, but rebounded to 70,000 by 2008. The reason for the drastic decline is unknown.

John Hoyt

Emperor Geese are seen rarely along the coast south of the Alaska panhandle. In British Columbia, Washington, and Oregon they are usually seen singly, but are also found in flocks of Brant, Snow Geese, Canada Geese, Cackling Geese, and Glaucous Gulls. They like rocky shores, breakwaters, jetties, and spits, but can be well camouflaged there. Geese with white heads and dark bodies should be checked carefully to distinguish Emperor Geese, which have a black throat and rump, and black and white scale marks on the light gray body, from blue-phase Snow Geese and Ross's Geese. Emperor Geese become unwary when associating with domestic Mallards (Campbell et al. 1990; Marshall et al. 2003; Mlodinow 2005).

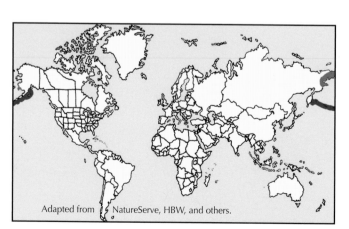

Adapted from NatureServe, HBW, and others.

NatureServe
Conservation Status

Global: .... Vulnerable

Canada: .... Not Applicable

US: .... Vulnerable

Mexico: .... Not Applicable

www.natureserve.org

Snow Goose

Scientific: *Chen caerulescens*
Français: Oie des neiges
México: Ganso Blanco
Order: Anseriformes
Family: Anatidae

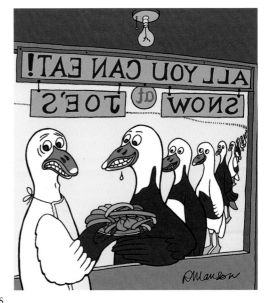

To grub means to dig up from the roots. Have you tried grubbing a dandelion, only to snap the root, thus leaving part of it in the ground and allowing the weed to grow again, much to your dismay? The Snow Goose's efficient grubbing of food plants has become a big problem in its northern breeding grounds. Its population has increased too much in the past decades for its breeding grounds to sustain it, because humans have been feeding it crops voluntarily or otherwise on its migration and wintering areas. All these surviving geese go back north and grub the coastal tundra, leaving only mud. The key tool this species employs for destroying its habitat is its specially shaped bill resembling a grinning patch that is formed by the exposed tooth-like serrations. It is a useful field mark and a reminder of the tight relationship that can exist between one species' morphology and ecosystem conservation.

The Snow Goose's two color morphs (white and blue) were once thought to be two species, but were really two species in the making. They were geographically separated on the wintering grounds and the breeding grounds. However, merging of the two morphs began in the early 1900s because of changing farming practices in the wintering areas that allowed the morphs to meet in rice-growing areas. Mixed pairs formed there and migrated together to the northern breeding grounds, thus halting the process of speciation.

Tom Whetten

Thus, the simple act of crop farming inadvertently led to a population explosion in an Arctic-breeding species, habitat destruction thousands of kilometers north of the abundant grain, and an end to this fine example of speciation (Mowbray et al. 2000).

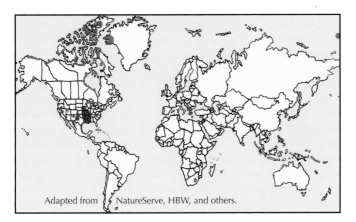

Adapted from NatureServe, HBW, and others.

NatureServe
Conservation Status

Global: .... Secure
Canada: .... Secure
US: .... Secure
Mexico: .... Not Applicable

www.natureserve.org

Ross's Goose

Scientific: *Chen rossii*
Français: Oie de Ross
México: Ganso de Ross
Order: Anseriformes
Family: Anatidae

All bird species are vulnerable to predation at some point in their life cycle. The Ross's Goose is the smallest goose that breeds in North America and it usually nests in colonies. It is too small to cope with some predators, especially the Arctic Fox (*Vulpes lagopus*). This may be responsible for its small breeding distribution. Nesting on islands in tundra lakes protects it from foxes, which often can't access islands. The inland location of the lakes reduces predation by gulls and jaegers, which are much more common along coasts. Most predation on Ross's Goose nests is by foxes, whereas only 2% is by birds. In some years, foxes cross ice bridges and destroy many nests, which can cause the rest to be abandoned, especially if this happens early in the nesting period. Nevertheless, the presence of the incubating female and the guarding male helps protect the nest to some degree. The female covers the eggs with down when leaving the nest. This helps maintain temperature and humidity and aids in concealing the eggs from predatory birds. Goslings flatten themselves on the ground to hide from predators. On water, they dive to elude predators. Both parents, especially the male, protect the young. Some dead adults are usually seen in nesting colonies by biologists, but, as in other birds, the number is likely much lower than the number of goslings killed, which in turn is likely lower than the number of pilfered eggs.

Edgar T. Jones

Ross's Geese may benefit by nesting close to Snow Geese, which are larger and better able to deter predators. Snow Geese are also more wary of predators during the hunting season, and this may be one reason why Ross's Geese often migrate with Snow Geese.

In 1930 fewer than 6,000 Ross's Geese were thought to exist in the world, but the population size has been increasing since the early 1950s such that, at perhaps one million individuals, this species is now another important factor in the decline of tundra habitat, along with the much more numerous Snow Goose (Ryder and Alisauskas 1995).

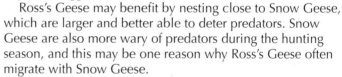

Adapted from NatureServe, HBW, and others.

NatureServe
Conservation Status

Global: ...... Apparently Secure

Canada: ..... Apparently Secure

US: .... Apparently Secure

Mexico: .... Not Applicable

www.natureserve.org

Brant

Scientific: ***Branta bernicla***
Français: Bernache cravant
México: Ganso de Collar
Order: Anseriformes
Family: Anatidae

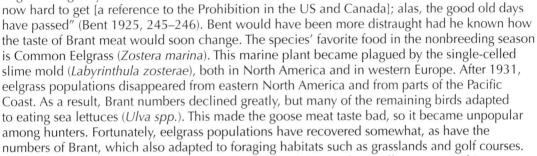

"From the standpoint of the epicure the Brant is one of our finest game birds, in my opinion *the* finest, not even excepting the far-famed Canvasback. I cannot think of any more delicious bird than a fat, young Brant, roasted just right and served hot with a bottle of good Burgundy. Both the bird and the bottle are now hard to get [a reference to the Prohibition in the US and Canada]; alas, the good old days have passed" (Bent 1925, 245–246). Bent would have been more distraught had he known how the taste of Brant meat would soon change. The species' favorite food in the nonbreeding season is Common Eelgrass (*Zostera marina*). This marine plant became plagued by the single-celled slime mold (*Labyrinthula zosterae*), both in North America and in western Europe. After 1931, eelgrass populations disappeared from eastern North America and from parts of the Pacific Coast. As a result, Brant numbers declined greatly, but many of the remaining birds adapted to eating sea lettuces (*Ulva spp.*). This made the goose meat taste bad, so it became unpopular among hunters. Fortunately, eelgrass populations have recovered somewhat, as have the numbers of Brant, which also adapted to foraging habitats such as grasslands and golf courses.

The Brant is a strong, agile flier. It attacks nest predators such as gulls, ravens, and jaegers that enter its breeding territory, and it often migrates in long nonstop flights. Flocks in flight give "the appearance of a series of regular and swift [vertical] waving motions such as pass along a pennant in a slight breeze" (Nelson 1881). They may also gracefully assemble themselves into a massive ball, then straighten out in an undulating formation.

Dennis Paulson

This species sometimes associates with diving ducks and American Coots, stealing food that they bring to the surface (see American Wigeon account, p. 26).

The Brant's well-developed salt gland allows it to drink salt water. However, it also drinks fresh water that flows into its marine feeding areas (Reed et al. 1998).

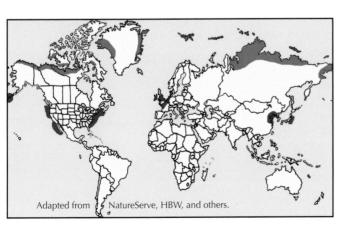

Adapted from NatureServe, HBW, and others.

NatureServe
Conservation Status

Global: .... Secure
Canada: ..... Apparently Secure
US: .... Secure
Mexico: .... Not Applicable

www.natureserve.org

Cackling Goose

Scientific: *Branta hutchinsii*
Français: Bernache de Hutchins
México: Ganso Canadiense Menor
Order: Anseriformes
Family: Anatidae

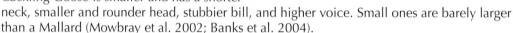

Approximately 12 subspecies of the Canada Goose have been recognized, but they were recently reorganized into two species based on differences in breeding range, migration-timing, size, voice, habitat, and genetics: five of them are now together considered the Cackling Goose, a "new" species. The other seven make up the new Canada Goose species. The tundra-nesting Cackling Goose is smaller and has a shorter neck, smaller and rounder head, stubbier bill, and higher voice. Small ones are barely larger than a Mallard (Mowbray et al. 2002; Banks et al. 2004).

Taxonomic changes are often met at first with frustration from birders and others who resist having to adapt. Many recent such changes stem largely from phylogenetic relationships being revealed by molecular genetics. But science often takes one step forward and one or two steps back. In fact, no guarantee exists that Canada Geese and Cackling Geese will always remain as two species, or that the 12 sub-species won't be re-organized into any number of species. It's not the stable world of mathematics, but the nature of nature.

Cackling Geese can be much smaller than Canada Geese.

In the birder whose life list grows by one species simply because of a species split, the chagrin eventually turns to elation. Since 2004, birders have been expending great effort during Canada Goose migration to find the needle in the haystack: one individual or a family group of Cackling Geese. The trick is to consider that much variation exists in all the field marks discussed above, but if one is patient and lucky, one may chance upon a convincing scene like the one at the left. Other possible haystack needles include White-fronted Geese and Brant.

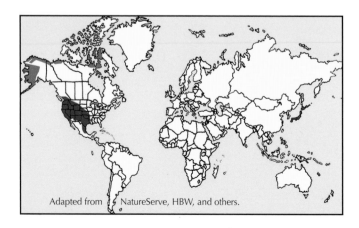

Adapted from NatureServe, HBW, and others.

NatureServe
Conservation Status

Global: .... Secure

Canada: .... Secure

US: .... Unrankable

Mexico: .... Unrankable

www.natureserve.org

Canada Goose

Scientific: ***Branta canadensis***
Français: Bernache du Canada
México: Ganso Canadiense Mayor
Order: Anseriformes
Family: Anatidae

"What man is so busy that he will not pause and look upward at the serried ranks of our grandest wildfowl, as their well-known honking notes announce their coming and their going, he knows not whence or whither? It is an impressive sight well worthy of his gaze; perhaps he will even stop to count the birds in the two long converging lines; he is sure to tell his friends about it, and perhaps it will even be published in the local paper as a harbinger of spring or a foreboding of winter" (Bent 1925, 204).

About two million Canada Geese are harvested annually in the US, with another 600,000 in Canada. The annual take has increased steadily since the 1970s, making the Canada Goose one of the three species of waterfowl most harvested in North America along with the Mallard and the Green-winged Teal (Mowbray et al. 2002). Many feral, free-flying, usually nonmigratory populations exist in the US and Canada—some are outside the species' normal breeding range.

The Canada Goose was introduced to Europe, initially in Britain around 1650. It occurs frequently in parks and is increasing its numbers and range rapidly on that continent, where it is now considered a pest (as in many parts of the US and Canada), mostly because of its droppings. Yet introductions to enhance hunting opportunities continued in Finland, Norway, and in parts of eastern Europe into the 1990s (Kirby and Sjöberg 1997). It was also introduced into New Zealand, Argentina, and Chile.

Dan Parent

Hunting is often the best way to control populations. Local problems at city parks, sport fields, and golf courses, where hunting and other lethal methods are usually not an option, are best handled with non-lethal means such as translocation, harassing (e.g., with a broom), scaring with trained dogs or noise-making devices, destroying eggs, or modifying the landscape to make nuisance sites and nearby feeding areas less attractive to Canada Geese (not "Canadian" geese).

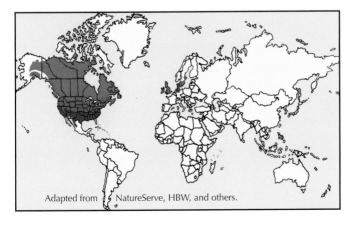

Adapted from NatureServe, HBW, and others.

NatureServe
Conservation Status

Global: .... Secure
Canada: .... Secure
US: .... Secure
Mexico: .... Not Applicable

www.natureserve.org

Trumpeter Swan

Scientific: *Cygnus buccinator*
Français: Cygne trompette
México: Cisne Trompetero
Order: Anseriformes
Family: Anatidae

"A swan seen at any time of the year in most parts of the US is the signal for every man with a gun to pursue it…. In the ages to come, like the call of the Whooping Crane…, the trumpetings that were once heard over the breadth of a great continent … will be locked in the silence of the past" (Forbush 1912, 475–476).

Perhaps one reason why Trumpeters decreased so much compared to the Tundra Swan was the Trumpeter's habit of migrating along the shore, where hunters could easily hide, whereas Tundras flew over large waters (Taverner 1974). Fortunately, programs to protect and reintroduce Trumpeter Swans have increased the adult population from dangerously low numbers in the early 1900s to 34,803 birds in 2005 (Moser 2006).

Despite its increasing population size and expanding breeding range, the Trumpeter Swan is still at risk from loss of wintering habitat, concentration of birds at few wintering sites, and, in some flocks, lack of migration. When it almost went extinct, knowledge of many of its migration routes and wintering sites, which parents had taught their cygnets, was lost. Current population size thus seems to be less important to its survival than the number of potential wintering sites and the birds' ability to learn to return to wintering sites year after year. The more the species is able to spread out in winter, the less vulnerable it is to overcrowding, disease, and harsh winter conditions at a particular site (Shea 1993). Birds from restored flocks have recently established new routes and increased their winter range (Mitchell and Eichholz 2010) by consulting their "migration and wintering archives", but the global population is still only Apparently Secure.

Robert Alvo

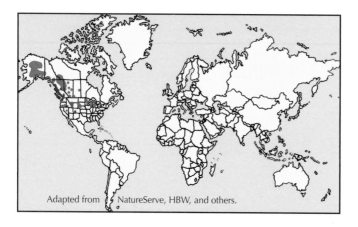

Adapted from NatureServe, HBW, and others.

NatureServe
Conservation Status

Global: ...... Apparently Secure

Canada: ...... Apparently Secure

US: .... Apparently Secure

Mexico: ...Presumed Extirpated

www.natureserve.org

Tundra Swan

Scientific: *Cygnus columbianus*
Français: Cygne siffleur
México: Cisne de Tundra
Order: Anseriformes
Family: Anatidae

In an Old World legend, swans never sang until just before death when they cried out a beautiful but sad song. This inspired the term "swan song", which means a person's final work, such as an artist's last painting. For example, after throwing down this book in disgust, some overly serious birders (think of the movie "The Big Year"), might proclaim that the long-term study that I published on the effects of acidification on Common Loon breeding success (see p. 81) was my swan song.

You can hear a Tundra Swan's powerful wings when it is 50 m above you. It uses them in battle, especially in territorial defense. A nesting bird may tackle a persistent intruding bird, beat it with its wings, peck it, stand on it, and, finally, trample it. In flight, one bird may chase the other and seize its tail, both birds falling toward the ground, after which they might or might not come out alive (Limpert and Earnst 1994).

Bent (1925, 289) told of Tundra Swans in the Niagara River in spring, where danger loomed on misty and foggy nights: "In one instance over 100 of these great birds met their death … in the Niagara swan trap.… Being caught in the rapids they were swept over the falls; many were killed by the fall, others were killed or maimed by the rough treatment they received in the whirlpools and rapids, where they were hurled against the rocks or crushed in the ice".

Bill Schmoker

Europeans call it the Bewick's Swan, whereas North Americans called it the "Whistling Swan" because of its distinctive vocal whistle. It stopped breeding in Manitoba, Ontario, and Quebec for over 150 years because of intense trade in skins and quills, but it has been returning.

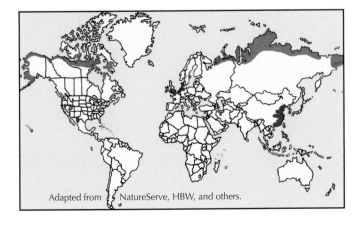

Adapted from NatureServe, HBW, and others.

NatureServe
Conservation Status

Global: .... Secure

Canada: .... Secure

US: .... Secure

Mexico: .... Not Applicable

www.natureserve.org

Wood Duck

Scientific: *Aix sponsa*
Français: Canard branchu
México: Pato Arcoiris
Order: Anseriformes
Family: Anatidae

A century ago, the way in which Wood Duck chicks traveled from the nest hole to the ground uninjured was a mystery. One theory held that the parents carried them down in their bills or on their backs. In fact, newly hatched young have sharp claws for climbing inside the tree from the nest up to the cavity entrance, up to four meters. They drop lightly down to the ground or water, 1–15 m below. Imagine being present at exactly the right moment to witness the magical event when the young jump out of the hole on their first adventure!

The Wood Duck has been called the "tree duck" from its habit of perching and nesting in trees. It nests in natural tree cavities or abandoned nest holes of large woodpeckers (especially Pileated Woodpeckers and, historically, Ivory-billed Woodpeckers). Nest boxes are often installed for them by hunting groups and other conservation organizations, but surprisingly it isn't clear that this boosts populations (Hepp and Bellrose 1995). Other ducks that nest regularly in tree cavities are the Black-bellied Whistling-Duck, Bufflehead, Common Goldeneye, Barrow's Goldeneye, Hooded Merganser, and Common Merganser.

American Beavers (*Castor canadensis*) were almost extirpated from North America by over-trapping. They help Wood Ducks by creating swamps through their dam- and dyke-construction.

Dan Parent

NatureServe
Conservation Status

Global: .... Secure

Canada: .... Secure

US: .... Secure

Mexico: .... Imperiled

www.natureserve.org

Adapted from NatureServe, HBW, and others.

Gadwall

Scientific: *Anas strepera*
Français: Canard chipeau
México: Pato Friso
Order: Anseriformes
Family: Anatidae

This is the first of our 10 dabbling ducks (genus *Anas*), most of which burst upward into flight instead of first pattering along the surface. They dabble, submerge their heads, and tip up rather than dive for food like diving ducks. *Chipeau*, the French name, means "whiner" (Donavan and Ouellet 1993), referring to the male's shrill whistled call.

Gadwalls eat floating and submerged vegetation, generally in deeper water than that in which other dabblers feed. This habit makes them less susceptible to body parasites than are many other ducks. Feeding in the water column rather than on the bottom makes their ingestion of spent lead shot from hunting rare. And by feeding in open-water habitats rather than in the wetland's fringe, they probably have been less affected than other dabblers by humans impinging on wetland edges. Indeed, the breeding range changed during the 1900s more than for any other duck in North America, expanding into eastern North America and west of the Cascade Mountains in the west (Leschack et al. 1997). The Gadwall has also expanded its range more than any other waterfowl has in Europe over the past 200 years (Berndt 1997). Perhaps the Gadwall's food niche is partly responsible for its greater expansion in North America and Europe in relation to other dabblers.

Gadwalls achieve higher nest success rates than other dabblers do because they often nest on islands to avoid mammalian predators. Also, they nest later in the season than do other ducks, at a time when they can better assess wetland quality, nest-site competition is reduced, and predation rates are low (Leschack et al. 1997).

A hen may push eggs to a new nest using the underside of her bill to avoid hindrances like traps set by bothersome biologists to catch hens on the nest (Blohm 1981).

Dennis Paulson

Robert Alvo

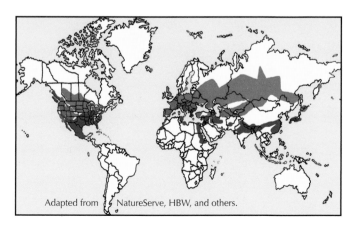

Adapted from NatureServe, HBW, and others.

NatureServe
Conservation Status

Global: .... Secure
Canada: .... Secure
US: .... Secure
Mexico: .... Imperiled

www.natureserve.org

American Wigeon

Scientific: *Anas americana*
Français: Canard d'Amérique
México: Pato Chalcuán
Order: Anseriformes
Family: Anatidae

The American Wigeon used to be called the "baldpate", referring to its conspicuous white forehead and crown. The male's soft *wid-wid-wigeon* whistle evokes its current name. This species is also called "poacher" for its habit of stealing plant food from birds that dive to the bottom to dig it up, notably Canvasbacks, Redheads, and American Coots. Fisher (1975) watched a female poaching plants from two small Common Muskrats (*Ondatra zibethicus*).

It is the most vegetarian of the regularly breeding North American dabblers, being almost exclusively so in winter and migration. It is well adapted to grazing on both aquatic and terrestrial plants. The wide and tall junction of the upper and lower mandibles increases the force it can exert at the bill tip to pluck firmly-rooted vegetation. Only females during early incubation and young ducklings take significant proportions of animal food (Mowbray 1999).

The American Wigeon's breeding range has been expanding eastward, perhaps because of the loss of habitat in the prairies and parklands of western North America, combined with the creation of more suitable habitat in eastern North America such as farm ponds, sewage lagoons, and impounded marshes. Eastward expansion should continue over the foreseeable future because much of this new habitat continues to be created under the auspices of the very successful North American Waterfowl Management Plan (Mowbray 1999).

Edgar T. Jones

Dennis Paulson

Birders can add a species to their list by closely scanning flocks of American Wigeons for lone Eurasian Wigeons, which will often oblige by staying apart from the flock (Floyd 2008). Eurasian Wigeons also are vegetarian thieves (Cramp 1977).

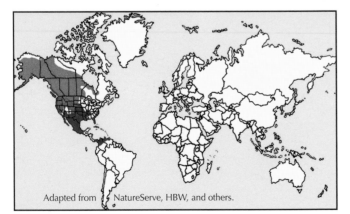

Adapted from NatureServe, HBW, and others.

NatureServe
Conservation Status

Global: .... Secure
Canada: .... Secure
US: .... Secure
Mexico: .... Not Applicable

www.natureserve.org

American Black Duck

Scientific: *Anas rubripes*
Français: Canard noir
México: Pato Sombrío
Order: Anseriformes
Family: Anatidae

"The American Black Duck is the wise king of all ducks [in the east, but see the Canvasback account, (p. 35) for the west's royal duck]. Anyone killing them consistently and in large numbers is probably cheating" (Walsh 1971, 11). Indeed, *The Outlaw Gunner*, an illustrated book by H. M. Walsh, reveals the wildfowler's secrets. These market hunters made their living selling thousands of waterfowl to markets after feeding their own families. In the late 1800s, they legally corn-baited to concentrate a region's ducks in one place for easy slaughter. "Tollers" were live waterfowl decoys trained to bring wild birds into shooting range. Simple duck traps caught dozens without wasting ammunition. Night "gunner lights" blinded waterfowl into submission. The gunner would lie hidden on his stomach in his skiff behind the light and gun fixed on the bow. "Big guns" were oversized shotguns that felled 100 ducks in one shot. Sinkboxes were wooden rafts with a central casket-like keel, in which the man could lie in wait, surrounded by decoys. His weight kept the raft at water surface level, making it invisible. One old-timer stated, "If you understand their feeding habits, you can make them sit on the end of your gun barrel" (Walsh 1971, 33). Once laws were passed, methods were modified into tricks to circumvent them. The outlaws hid killed ducks in hollow decoys or they took ducks from traps alive in a bag that could be opened on short notice to allow the evidence to fly away. Once most of the big guns had been confiscated, boiler pipe was used to make battery guns (e.g., seven barrels lying side by side on the bow). The outlaws also set duck

Julie Dufour

traps as decoys so that after the warden destroyed them and declared the area "clean", more traps would be set.

The American Black Duck's Achilles' heel in those days was its love of corn combined with its habit of feeding at night. Now that laws in the US and Canada emanating from the 1916 Migratory Bird Convention are both respected and enforced, this species' main problems are competition and hybridization with the Mallard and overhunting, but the Black Duck Joint Venture is keeping a close eye on it as part of the North American Waterfowl Management Plan (NAWMP).

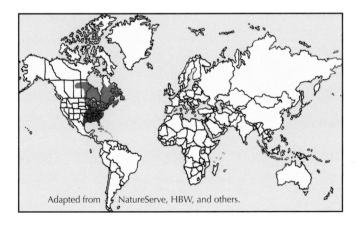

Adapted from NatureServe, HBW, and others.

NatureServe
Conservation Status

Global: .... Secure
Canada: .... Secure
US: .... Secure
Mexico: .... Not Applicable

www.natureserve.org

Mallard

Scientific: *Anas platyrhynchos*
Français: Canard colvert
México: Pato de Collar
Order: Anseriformes
Family: Anatidae

At 19 million breeding birds, the Mallard is one of the world's best-known waterfowl. It is the most widely distributed, most abundant, and most heavily hunted duck in North America, where its population has cycled from 5 million to 11 million since 1955. Modern waterfowl management here is currently based largely on what is known about this species. It can thrive in a wide range of landscapes. Often in close contact with humans, it is the source of all domestic ducks except the Muscovy Duck. It has been bred selectively to maximize meat or egg production, to be used as live decoys, or to experiment with colorful plumages. Many shapes, sizes, and color combinations have thus been created. Feral populations composed of mixtures of wild Mallards and domesticated breeds live in urban areas throughout the world, where they habituate to humans who feed them. Mallards hybridize with closely related species (e.g., American Black Ducks and Mottled Ducks in North America), thus reducing the pure quality of those species' genetic identity (Drilling et al. 2002).

Males, with their unique upward-curving black central tail-feathers, abandon their incubating mates and gather in small flocks to molt into eclipse (concealing) plumage. Unlike many other birds, which wear a breeding plumage in spring and summer and a nonbreeding plumage in fall and winter, ducks wear a breeding plumage in winter and spring and an eclipse plumage for a short time in summer. Given their high weight/wing surface-area ratio, the loss of only a few feathers would make them flightless, so it makes sense to shed them all at once to reduce the flightless period, during which they hide in emergent vegetation. Aquatic birds such as waterfowl, loons, grebes, anhingas, and even dippers can tolerate being flightless during molt because they can flee from predators by diving. Many terrestrial birds would be in big trouble if rendered flightless (Podulka et al. 2004).

Female Mallards make the familiar descending "quack".

Peter Sproule

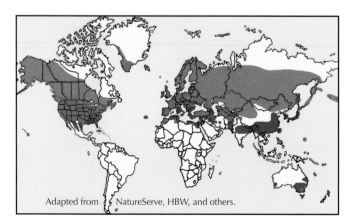

Adapted from NatureServe, HBW, and others.

NatureServe
Conservation Status

Global: .... Secure

Canada: .... Secure

US: .... Secure

Mexico: .... Secure

www.natureserve.org

Mottled Duck

Scientific: *Anas fulvigula*
Français: Canard brun
México: Pato Tejano
Order: Anseriformes
Family: Anatidae

This is the only nonmigratory waterfowl species regularly breeding in North America. Its maximum known movement is only 430 km, a rare response to major changes in habitat conditions (Bielefeld et al. 2010).

Extreme recorded dates show that its breeding season is very long: a new nest found in early February, and a brood seen in late December. Much variation in timing can occur from year to year depending on water levels, with peak nest initiation dates varying by more than 60 days. Low water levels in late winter and early spring often delay nesting (Bielefeld et al. 2010).

Hybridization with Mallards is the Mottled Duck's main threat (it is Apparently Secure globally and nationally in the US and Mexico). The Western Gulf Coast (WGC) population has 210,000–630,000 birds in spring, whereas the Florida population has only 28,000–53,000. Many of the Florida birds are actually Mallards or hybrids, all these "brown ducks" being difficult to distinguish in surveys. Information on relative percentages of the different "brown ducks" in the population is badly needed. A complication is that Mottled Duck populations in South Carolina and Georgia, a result of introductions of WGC birds, have spread to Florida. The WGC population has a low hybridization rate, which should be monitored because it could easily increase. Releases of domestic Mallards must be strongly discouraged in the US and all over the world because they threaten other closely related forms, such as New Zealand's Grey Duck (*Anas s. superciliosa*), through hybridization. Mottled Duck numbers in Mexico are thought to be small despite the low monitoring intensity.

Our least social dabbler, the Mottled Duck is usually found in pairs or in flocks of fewer than 10 individuals (Bielefeld et al. 2010).

Larry Master

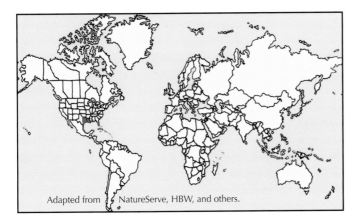

Adapted from NatureServe, HBW, and others.

NatureServe
Conservation Status

Global: ...... Apparently Secure

Canada: .... Not Applicable

US: .... Apparently Secure

Mexico: ..... Apparently Secure

www.natureserve.org

Blue-winged Teal

Scientific: *Anas discors*
Français: Sarcelle à ailes bleues
México: Cerceta Alas Azules
Order: Anseriformes
Family: Anatidae

"There beneath a tuft of grasses in a hollow on the ground was the nest built of grasses and lined with dark-brown mottled down pulled from the mother's own breast. In the midst of the downy bedclothes rested 10 beautiful, cream-colored eggs—an exquisite casket of jewels destined to develop into living gems far lovelier than any rubies or diamonds ever dug from the earth.... Two weeks later we found the nest empty, but the whole family were out there on the pond bobbing about as buoyant as corks, learning how to make a living and survive in a wonderful but dangerous world" (Townsend 1910). People have long been fascinated by wild bird eggs. In the 1800s, collecting them was a natural part of growing up in North America, particularly for boys, but the hobby is now illegal. Great Britain used to have a periodical called *The Oologist* that was devoted to the study, collection, and trade of bird eggs and nests.

In the 1990s, the Blue-winged Teal replaced the Lesser Scaup as the second most abundant duck in North America after the Mallard (Drilling et al. 2002). As with other waterfowl, its population fluctuates greatly with wetland conditions in the prairies, where rain incites people to exclaim "good weather for the ducks!" The North American Waterfowl Management Plan's (NAWMP) population goal is 4.7 million breeders, but numbers have fluctuated between 2.8 million and 7.4 million since standardized surveys began in 1955. Mammalian predators often take 90% of nests. An estimated 900,000 adults (75% females) are taken by Red Foxes

Dan Parent

(*Vulpes vulpes*) alone each year in the prairie-pothole region. Predation elsewhere in its range and by other predators, combined with an annual take by hunters in the hundreds of thousands if not in the millions (numbers are lacking for most of its wintering range, which is in Latin America), must also be taken into account by population managers (Rohwer et al. 2002). It is these large fluctuations that demand careful management of hunted species, even the Secure Blue-winged Teal.

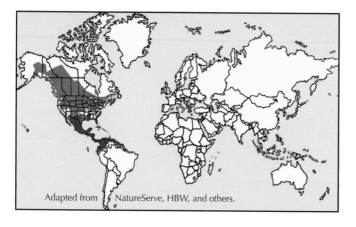

Adapted from NatureServe, HBW, and others.

NatureServe
Conservation Status

Global: .... Secure

Canada: .... Secure

US: .... Secure

Mexico: .... Not Applicable

www.natureserve.org

Cinnamon Teal

Scientific: *Anas cyanoptera*
Français: Sarcelle cannelle
México: Cerceta Canela
Order: Anseriformes
Family: Anatidae

The name *cyanoptera* refers to the blue wing patch, which the Cinnamon Teal shares with the Blue-winged Teal and the Northern Shoveler. The Cinnamon Teal is the most western-ranging of the North American ducks, its center of abundance being situated west of the Rocky Mountains. Unlike most North American dabbling ducks, however, it only rarely breeds in the prairie and parkland regions.

The Cinnamon Teal nests closer to water than other dabbling ducks do, usually within 50 m, and often within one meter. The hen approaches and leaves the nest by walking as much as 70 m through vegetation, often through tunnels. Like the closely related Northern Shoveler, it engages in social feeding in which groups of birds follow each other, dabbling in the water stirred up by the bird in front (Gammonley 2012).

It is one of the least studied waterfowl species in North America. Good population estimates are lacking because counts often include Blue-winged Teals, and much of its core breeding range lies south of the standardized waterfowl survey area on the northern prairies, for example in Nevada (Floyd et al. 2007). Bellrose's estimate of 260,000–300,000 breeders in the 1970s indicates that it is one of the least abundant dabbling ducks in North America. Long-term population trends also are poorly known. Its early fall migration before hunting season makes it only lightly harvested in Canada and the United States, and harvest in Mexico and the neotropics also is light. Habitat availability is the key limiting factor to its population in the arid west.

Steve Zamek

Competition for resources with the Blue-winged Teal may limit expansion of the Cinnamon Teal's breeding range because this species is subordinate to the Blue-winged Teal (the latter initiated and won most observed encounters between the two). However, the Cinnamon Teal is better suited for wetland landscapes in the west (Gammonley 2012) and deserves the title of Western Champ.

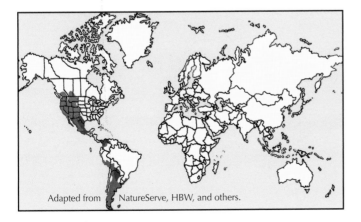

Adapted from NatureServe, HBW, and others.

NatureServe
Conservation Status

Global: .... Secure

Canada: ..... Apparently Secure

US: .... Secure

Mexico: .... Secure

www.natureserve.org

Northern Shoveler

Scientific: *Anas clypeata*
Français: Canard souchet
México: Pato Cucharón Norteño
Order: Anseriformes
Family: Anatidae

The species name *clypeata*, meaning having a shield, refers to the Northern Shoveler's huge bill that is distinctive even in two-week-old ducklings. Many comb-like "teeth" furnish the sides of the bill, straining small invertebrates and seeds while the bird paddles along the surface. Unlike other dabbling ducks, Northern Shovelers don't feed on land. At take-off, this species' wings make a characteristic rattling sound that likely plays an advertisement role during the breeding season (Dubowy 1996).

Male Northern Shovelers are the most territorial of all North American dabbling ducks, remaining with mates longer than the others do. Another unique characteristic among waterfowl: males don't undertake an extensive molt migration. Rather, they molt in the breeding region in small flocks. Once they actually become flightless, making them more vulnerable to predators, they are solitary, hide in emergent vegetation day and night, and feed very little, whereas other dabbling ducks remain in open water or are active at night (Dubowy 1996).

Juan Bahamon

Robert Alvo

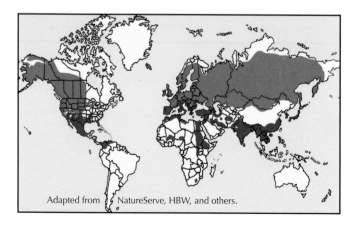

Adapted from NatureServe, HBW, and others.

NatureServe
Conservation Status

Global: .... Secure

Canada: .... Secure

US: .... Secure

Mexico: .... Not Applicable

www.natureserve.org

Northern Pintail

Scientific: *Anas acuta*
Français: Canard pilet
México: Pato Golondrino
Order: Anseriformes
Family: Anatidae

"Vying with the Mallard to be the first of the surface-feeding ducks to push northward on the heels of retreating winter, this hardy pioneer extends its migration to the Arctic coast of the continent and occupies the widest breeding range of any North American duck" (Bent 1923, 144). It is the most abundant Arctic-breeding duck (Suchy and Anderson 1987). Its pointed look is striking. Note the slim overall shape that is accented by the long neck, the pointed tail (even in the female), and the breeding male's fine white neckline.

This is the only North American dabbling duck that prefers nesting in tilled cropland (mainly grain stubble). Unfortunately, these nests are often initiated early in the nesting season and are later destroyed in large numbers by crop cultivation and planting. Northern Pintails tend to nest farther from water than do other ducks, sometimes 3 km away (Austin and Miller 1995).

(a)

(b)

(c)

(d)

PHOTOGRAPHERS:
(a) Bill Schmoker
(b) Peter Sproule
(c) Darroch Whitaker
(d) Robert Alvo

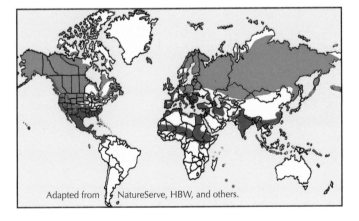

Adapted from NatureServe, HBW, and others.

NatureServe
Conservation Status

Global: .... Secure
Canada: .... Secure
US: .... Secure
Mexico: .... Not Applicable

www.natureserve.org

Green-winged Teal

Scientific: *Anas crecca*
Français: Sarcelle d'hiver
México: Cerceta Alas Verdes
Order: Anseriformes
Family: Anatidae

"Nothing about its rankness of flavor when it has gorged on putrid salmon lying in the creeks in the northwest, or the maggots they contain, ever creeps into the books; and yet this dainty little exquisite of the southern rice fields has a voracious appetite worthy of the Mallard around the salmon canneries of British Columbia where the stench from a flock of teals passing overhead betrays a taste for high living no other gourmand can approve. When clean fed, however, there is no better table-duck than a teal" (Blanchan 1899, 103).

The Green-winged Teal, our smallest breeding dabbling duck, is the one most highly associated with wetlands surrounded by trees (Wood Ducks aren't considered true dabblers), and this has saved it from prairie droughts and intensive agriculture. It is usually the second most commonly hunted duck after the Mallard. Hunters know it for its large flocks, high flight speed, and aerial maneuverability. It is the only North American duck known to scratch itself in flight, scratching its bill with its feet, possibly to remove leeches from its nasal passages (Johnson 1995).

Its Old World counterpart, the Eurasian Teal, used to be considered a separate species, and it is again being treated as such in some Eurasian field guides (e.g., Brazil 2009). In anticipation of a possible re-splitting of the two forms here, we took great interest in comparing them in Alaska, for the easiest species you can add to your life list is one that you've already identified

Dennis Paulson

in the field as being a different form, then check off years later in the comfort of your home once receiving the good news in the latest supplement to the *Check-List of North American Birds* (American Ornithologists' Union 1998). The main difference in breeding males is the horizontal white line in the sitting Old World bird instead of the vertical white line in the New World one.

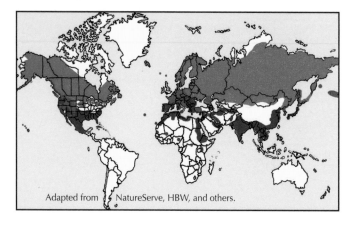

Adapted from NatureServe, HBW, and others.

NatureServe
Conservation Status

Global: .... Secure
Canada: .... Secure
US: .... Secure
Mexico: .... Not Applicable

www.natureserve.org

Canvasback

Scientific: *Aythya valisineria*
Français: Fuligule à dos blanc
México: Pato Coacoxtle
Order: Anseriformes
Family: Anatidae

Bent (1923, 189) may not have referred to this species as "lordly" had he understood that nesting females are duped by female Redheads into caring for young that aren't their own (see next account). He knew that wigeons and coots pirated Wild Celery (*Vallisneria americana*) from the Canvasback, but he assigned the duck top royalty status for western North America nonetheless based on its excellent taste, the challenge of killing it (it flies high and fast, and is hard to kill), and its sloped facial profile ("Roman nose"). Wild Celery, the Canvasback's favorite food in winter and on migration, made the bird a delicacy. Canvasbacks have changed their migration routes, wintering sites, and diets, largely because of a decline in Wild Celery (Mowbray 2002a).

In 1938, during the Dust Bowl, industrialist and hunter James Ford Bell was worried about declining North American waterfowl populations, so he provided funding and his hunting camp at Delta Marsh, Manitoba, to ecologist Aldo Leopold's student, Al Hochbaum, to set up a research station. Hochbaum did so and made many findings, including the discovery of the Canvasback's main nesting area in the prairie pothole ponds west of Delta. Many researchers at what later became Delta Waterfowl and Wetlands Research Station produced hundreds of waterfowl and wetland papers, including long-term studies that provided a unique synthesis of Canvasback biology (Delta Waterfowl 2011). I assisted in one of the long-term Canvasback studies, and fondly recall "drive-trapping" with other wetsuit-clad assistants. We would invade prairie potholes to capture broods of large flightless young in a funnel-shaped net for measurement and banding. It was fascinating to grope the mucky pond benthos around nests to identify eggs that had been displaced during scuffles between female Canvasbacks and parasitizing female Redheads.

bryanjsmith

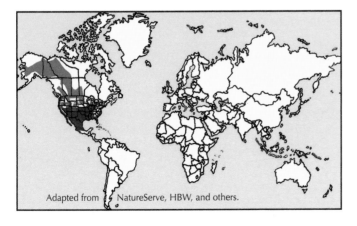

Adapted from NatureServe, HBW, and others.

NatureServe
Conservation Status

Global: .... Secure
Canada: .... Secure
US: .... Secure
Mexico: .... Not Applicable

www.natureserve.org

Redhead

Scientific: *Aythya americana*
Français: Fuligule à tête rouge
México: Pato Cabeza Roja
Order: Anseriformes
Family: Anatidae

"All of the slough-nesting ducks seem to be very careless about laying their eggs in the nests of other species" (Bent 1923, 193). Poor Bent would be kicking himself now, for we know he was at the cusp of understanding a sneaky game that has been going on every spring in perhaps the least appreciated, yet very rich, North American habitat. Redhead females are actually very careful to lay their eggs in nests of other ducks, especially Canvasbacks, to allow the latter to raise the yellow Redhead ducklings along with their own brown ones. The trick is called "non-obligate brood parasitism". Unlike Brown-headed Cowbirds (*Molothrus ater*), which lay their eggs only in the nests of other species, Redheads aren't obliged to. They lay eggs in their own nests too. Other aquatic bird species are also involved in this battle in North America's "duck factory" (www.ducks.org), as parasites, as hosts, or both.

Flying over the prairies' monotonous agricultural fields, look closely to see the mosaic of variously shaped gems dotting them—prairie potholes, each with its ring of emergent vegetation in which the diving ducks, grebes, coots, and gallinules nest, while the dabbling ducks use the adjacent fields. The broods of all meet in the potholes. With such a high density of aquatic birds sharing this bustling ecosystem, it isn't surprising that so much brood parasitism occurs there.

Right: This female Lesser Scaup on a prairie pothole is raising three of her own chicks (brown) and three Redhead chicks (yellow). Redhead females may parasitize nests of other ducks or lay eggs in their own nest, or both. This strategy is called "non-obligate nest parasitism".

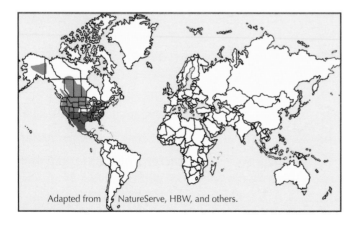

Adapted from NatureServe, HBW, and others.

Ring-necked Duck

Scientific: *Aythya collaris*
Français: Fuligule à collier
México: Pato Pico Anillado
Order: Anseriformes
Family: Anatidae

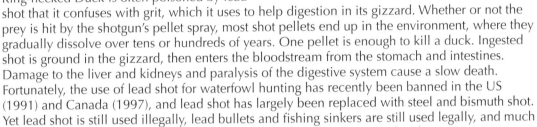

This species might better be called the "Ring-*billed* Duck", yet of the four names given above, only the Mexican one features the bill. The others refer to the cinnamon-colored ring at the base of the neck, a poor field mark seen only at close range and in good light.

As are many other waterfowl species that pick up food from aquatic sediments, the Ring-necked Duck is often poisoned by lead shot that it confuses with grit, which it uses to help digestion in its gizzard. Whether or not the prey is hit by the shotgun's pellet spray, most shot pellets end up in the environment, where they gradually dissolve over tens or hundreds of years. One pellet is enough to kill a duck. Ingested shot is ground in the gizzard, then enters the bloodstream from the stomach and intestines. Damage to the liver and kidneys and paralysis of the digestive system cause a slow death. Fortunately, the use of lead shot for waterfowl hunting has recently been banned in the US (1991) and Canada (1997), and lead shot has largely been replaced with steel and bismuth shot. Yet lead shot is still used illegally, lead bullets and fishing sinkers are still used legally, and much lead shot has already accumulated in the environment.

Larry Master

Many kinds of birds are still poisoned. For example, raptors and scavengers die when their prey contains lead (see California Condor account, p. 128). The presence of lead in the environment also poses a risk to humans through livestock contamination and human consumption of game taken with lead shot (Schuehammer and Norris 1995).

The Ring-necked Duck's generalized feeding habits compared to other *Aythya* diving ducks allows it to colonize areas in marginal waterfowl habitats such as bogs. Indeed, the species' breeding distribution expanded and its numbers increased during the 1980s and early 1990s when other North American ducks declined because of drought in the prairies (Roy et al. 2012).

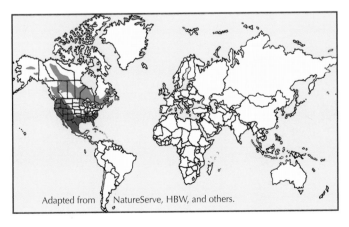

Adapted from NatureServe, HBW, and others.

NatureServe
Conservation Status

Global: .... Secure
Canada: .... Secure
US: .... Secure
Mexico: .... Not Applicable

www.natureserve.org

Greater Scaup

Scientific: *Aythya marila*
Français: Fuligule milouinan
México: Pato Boludo Mayor
Order: Anseriformes
Family: Anatidae

Why has the North American "bluebill" (Greater Scaup and Lesser Scaup combined) breeding population declined steadily from 6.3 million birds in 1980 to 3 million in 2006 despite reductions in hunting limits that started in 1986? No one knows yet. Several possible mechanisms have been or are being examined: 1) the ingestion of harmful amounts of selenium, especially on the Great Lakes, where introduced Zebra Mussels (*Dreissena polymorpha*) take in large amounts of the metal by filtering water before being eaten by scaup in large quantities; 2) high female mortality rates; 3) low breeding success rates; 4) habitat loss in the upper midwestern US spring migration stopover region, causing poor body condition; 5) an internal parasite introduced into the upper Mississippi River system, which caused repeated scaup die-offs during migration; 6) the disappearance of wetlands used for breeding, caused by climate change-induced loss of the permafrost seal that keeps water from draining out of northern wetlands; and, 7) changes in food availability in wetlands used for breeding, due to climate change. The Scaup Action Team acting under the North American Waterfowl Management Plan is focused on this problem because more research is clearly needed. The notion that water intake structures covered with Zebra Mussels attract foraging scaup that get sucked in and drown may provide fodder for interesting imagery, but little evidence exists to suggest that herein lies the answer to the mysterious decline of scaup.

Larry Master

A curious fact is that the Greater Scaup demonstrates almost factory-level consistency in the number of eggs in a clutch and in egg size. Its eggs can be distinguished from those of the Lesser Scaup, which, not surprisingly, tend to be a little smaller. Even though eggs of the two species may overlap slightly in one measurement (length, breadth, or mass), if all three readings are taken, the eggs of the two species can be easily separated. The typical clutch size of the Greater Scaup is eight or nine eggs (Kessel et al. 2002).

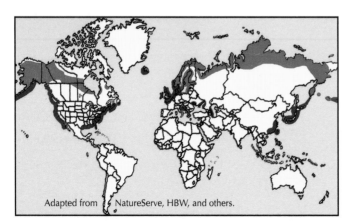

Adapted from NatureServe, HBW, and others.

NatureServe
Conservation Status

Global: .... Secure

Canada: .... Secure

US: .... Secure

Mexico: .... Not Applicable

www.natureserve.org

Lesser Scaup

Scientific: *Aythya affinis*
Français: Petit Fuligule
México: Pato Boludo Menor
Order: Anseriformes
Family: Anatidae

I asked Quebec biologist Michel Robert how to distinguish between the Lesser and Greater Scaups. "Head shape", he said in French. "The Greater's head looks round, whereas the Lesser's looks like it's been whacked from behind." Kevin Wallace nicely portrays the image that has remained fixed in my memory for more than 20 years. But Sibley (2000) notes that this field mark is best used when the birds are relaxed and not actively diving. The Lesser Scaup is smaller than the Greater Scaup, has a narrower neck and bill with smaller black bill tip, and has darker sides and flanks. In flight, the Greater shows a longer white wing stripe. The two species' similarity and the presence of nonbreeders in many breeding areas make it difficult to monitor populations separately. The Greater has a more northern distribution in all seasons, is more of a tundra duck, and also breeds in Eurasia, whereas the Lesser is more of a prairie and boreal forest bird that breeds only in the New World. In winter, the Greater prefers more open salt water, whereas the Lesser prefers fresh water and smaller ponds.

The Lesser Scaup takes longer migrating to its breeding grounds than do other North American ducks, and it pairs up later than they do, but once paired, females at different latitudes all nest around the same date. It is also the most mobile North American *Aythya* diving duck on land, often nesting a distance away from wetlands in so-called "uplands". Yet it is an excellent swimmer and diver, comfortable in rough water and large waves, and has been spotted 45–95 km offshore in the Gulf of Mexico (Austin et al. 1998).

Larry Master

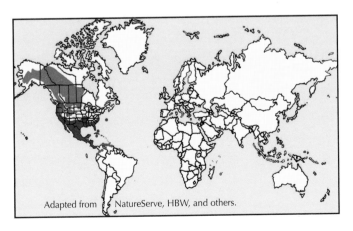

Adapted from NatureServe, HBW, and others.

NatureServe
Conservation Status

Global: .... Secure
Canada: .... Secure
US: .... Secure
Mexico: .... Not Applicable

www.natureserve.org

Steller's Eider

Scientific: *Polysticta stelleri*
Français: Eider de Steller
México: Eider de Steller
Order: Anseriformes
Family: Anatidae

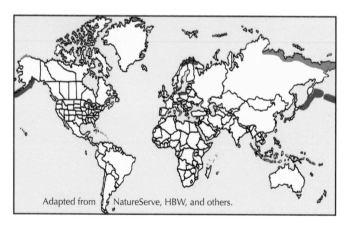

It is surprising that no one seems to have studied the question of why the breeding male Steller's Eider has four pairs of oddly placed spots. After all, it is the sort of puzzle for which some folks love to find fascinating explanations. The genus name *Polysticta* indeed means "many spots". An interesting aspect of these spots is that they all appear different. Viewed from the bird's side, one spot looks like a green scratch between the bill and eye, the second is a black eye, the third is a gray-green wound on the back of the head, and the last is a bullet hole on the flank. (As if having a premonition, hatchlings often show four light spots on the back, though these are gone in older downy ducklings.) Altogether, these spots give the bird a battered look symbolic of its Vulnerable global status, which reflects the species' now "spotty" breeding global range, its precipitous population decline from as-yet-unknown causes, and its current small population. Overharvesting, lead poisoning, oil spills, disturbance, increased predation, breeding habitat loss, and entrapment in fishing gear are possible factors.

The oddball of the world's four eider species, the Steller's Eider has its own genus. It has blue-and-white wing speculums in both sexes, is much smaller than the three other species,

Dennis Paulson

lacks bill-feathering in all plumages, is more graceful on land, lays more eggs on average (6–8 compared to 4–6), uses the littoral zone in winter rather than deeper water, and is a more generalized feeder. Males don't make a *coo* call and they use a rather different repertoire of courtship behaviors, while females use a more ritualized set of actions to incite males to court them than do the three larger eiders (Fredrickson 2001).

Adapted from NatureServe, HBW, and others.

NatureServe
Conservation Status

Global: .... Vulnerable

Canada: .... Not Applicable

US: .... Imperiled

Mexico: .... Not Applicable

www.natureserve.org

Spectacled Eider

Scientific: *Somateria fischeri*
Français:　Eider à lunettes
México:　Eider de Anteojos
Order:　Anseriformes
Family:　Anatidae

This species and the Steller's Eider have a very similar conservation status. Their breeding ranges are largely restricted to the northern coasts of Alaska and eastern Russia, and their global populations consist of only a few hundred thousand individuals. Furthermore, both species have recently suffered strong downward population trends, and they face similar threats. The Spectacled Eider has a larger population than the Steller's, and its decline seems to have slowed recently, but the fact that the global population spends most of the year in an exceptionally small area (50 × 75 km) makes it vulnerable to one site-specific disaster such as an oil spill. Indeed, one of the last great mysteries of North American bird distribution was solved in 1993 by implanting battery-powered satellite transmitters in some of the birds—the Spectacled Eider's main wintering area was found in the northern Bering Sea south of St. Lawrence Island (Petersen et al. 1999).

That island, purchased by the US from the Russian Empire for US$7.2 million in 1867 as part of "Seward's Folly", is the meeting place of the New and Old Worlds. (Locals told us that they visit relatives by motor boat, no passport required, on the opposite side of the Bering Sea, about 30 km away.) The New and Old World breeding populations of the Spectacled Eider, two in Alaska and one in Russia, are relatively close to each other geographically. No subspecies are known, but little mixing occurs now between the breeding populations despite the restricted wintering range.

In all plumages, including the downy one, this species has a large, round patch of feathers about the eye, as reflected in its three common names listed above. Its "spectacles" easily distinguish it from all other ducks (Petersen et al. 2000).

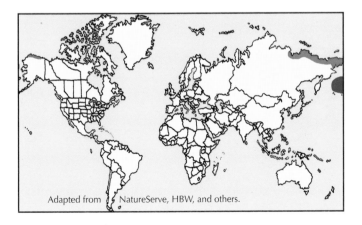

Dennis Paulson

Adapted from NatureServe, HBW, and others.

NatureServe
Conservation Status

Global: .... Vulnerable

Canada: .... Not Applicable

US: .... Imperiled

Mexico: .... Not Applicable

www.natureserve.org

King Eider

Scientific: *Somateria spectabilis*
Français: Eider à tête grise
México: Eider Real
Order: Anseriformes
Family: Anatidae

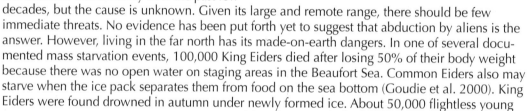

If you were an alien visiting from another world, and your boss tasked you with selecting one life form to abduct for her "best specimens from each planet" menagerie, which species would you select? An orchid, a butterfly, a toad? How about the King Eider? The species name *spectabilis* means "spectacular". Indeed, upon seeing "a male in full breeding finery, the most jaded birder's heart may skip a beat" (Koes 2003, 117).

The North American King Eider population has declined considerably over the past few decades, but the cause is unknown. Given its large and remote range, there should be few immediate threats. No evidence has been put forth yet to suggest that abduction by aliens is the answer. However, living in the far north has its made-on-earth dangers. In one of several documented mass starvation events, 100,000 King Eiders died after losing 50% of their body weight because there was no open water on staging areas in the Beaufort Sea. Common Eiders also may starve when the ice pack separates them from food on the sea bottom (Goudie et al. 2000). King Eiders were found drowned in autumn under newly formed ice. About 50,000 flightless young and females in molt died during a sudden freeze or epidemic on Banks Island. One autumn, 110 dead birds with evidence of trauma were found far from the ocean on Baffin Island. Ice fog was thought to have covered them with ice, causing them to crash-land. Evidence included eye damage that could have been caused by the freeze (Powell and Suydam 2012). To what extent were these problems associated with less-predictable weather and ice patterns due to human-induced climate change? We don't know, but scientists indeed forewarned us of less predictability and more extremes in weather some decades ago.

Edgar T. Jones

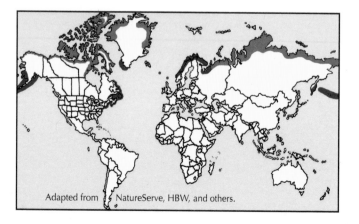

Adapted from NatureServe, HBW, and others.

NatureServe
Conservation Status

Global: .... Secure

Canada: .... Vulnerable

US: .... Apparently Secure

Mexico: .... Not Applicable

www.natureserve.org

Common Eider

Scientific: *Somateria mollissima*
Français: Eider à duvet
México: Eider Común
Order: Anseriformes
Family: Anatidae

"There is no reason why the Common Eider, which furnishes the valuable eider-down of commerce, should not be made a source of considerable income without any reduction of its natural abundance. The principle of conservation can as well be applied to the eider as to a forest" (Townsend 1914, 15–16). In Iceland, over the past thousand years, eiders "have become almost domesticated and are found in vast multitudes, as the young remain and breed in the place of their birth" (Baird et al. 1884, 74) as opposed to migrating south. In Quebec, however, conservationists have gone one step further. La Société Duvetnor Ltée. has been collecting down from eiders on islands in the St. Lawrence Estuary and using the proceeds to continue protecting the islands. Thus by giving up their down, at no detriment to themselves or to their reproductive success, females are essentially paying the rent for the population. This partnership model has been proposed for other parts of this eider's breeding range. The species name *mollissima* means "very soft", referring to the down that females use to line their nests. It is one of the warmest animal materials known, yet it is remarkably light, being used in sleeping bags, jackets, and pillows. It allows humidity to escape, but keeps heat in, and it bounces back to shape after being compressed (Bédard et al. 2008).

Nicole Bouglouan

European male. North American males might look slightly different (e.g., bill color).

Emily Pipher

Nests containing more than the average six eggs are usually "dump" nests involving more than one female.

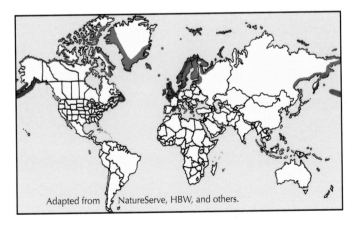

Adapted from NatureServe, HBW, and others.

NatureServe
Conservation Status

Global: .... Secure

Canada: .... Secure

US: .... Secure

Mexico: .... Not Applicable

www.natureserve.org

Harlequin Duck

Scientific: *Histrionicus histrionicus*
Français: Arlequin plongeur
México: Pato Arlequín
Order: Anseriformes
Family: Anatidae

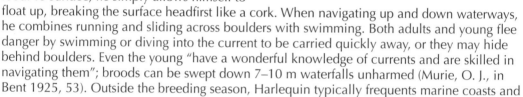

"Harlequin" was a stock character of 16th-century Italian comedy. In his avian form he specializes in breeding along clear, fast-flowing rivers and streams, where he eats invertebrates and fish roe by diving for them. The feet propel him down while he holds his wings slightly outstretched for steering or balance. He forages on the river bed by running with great agility against strong currents, his head held low as he pokes for food. To surface, he simply allows himself to float up, breaking the surface headfirst like a cork. When navigating up and down waterways, he combines running and sliding across boulders with swimming. Both adults and young flee danger by swimming or diving into the current to be carried quickly away, or they may hide behind boulders. Even the young "have a wonderful knowledge of currents and are skilled in navigating them"; broods can be swept down 7–10 m waterfalls unharmed (Murie, O. J., in Bent 1925, 53). Outside the breeding season, Harlequin typically frequents marine coasts and islands, often the roughest and rockiest shores where the water is churning and the surf breaks. The meaning of "actor" in both his genus and species name *histrionicus* could be taken to refer to his wonderful means of locomotion described above, but in fact refers to the male's head design when in breeding plumage.

Darroch Whitaker

Usually within five meters of a stream, nests may be located on the ground, on small cliff ledges, in tree cavities, or on stumps. Ground nests tend to be located on low-lying midstream islands. Re-nesting by female Harlequins that lose clutches to predators has surprisingly not been documented (Robertson and Goudie 1999).

Adapted from NatureServe, HBW, and others.

NatureServe
Conservation Status

Global: ...... Apparently Secure
Canada: ..... Apparently Secure
US: .... Apparently Secure
Mexico: .... Not Applicable

www.natureserve.org

Labrador Duck

Scientific: *Camptorhynchus labradorius*
Français: Eider du Labrador
México: Pato de Labrador
Order: Anseriformes
Family: Anatidae

L ittle is known about this sea duck that went extinct in the late 1800s. Market hunting was partly to blame, but ironically the poor-tasting birds often didn't sell. No definite breeding records exist, but the species may have nested on rocky nearshore islands. In winter it seems to have foraged mostly in shallow seawater, preferring a sandy bottom close to sheltered shores. Its flap-edged bill and the prominent vertical "teeth" in the lower mandible may have been used to sift mussels, clams, crayfish and/or small fish. The species may have been on the brink of extinction for natural reasons (e.g., specialized diet, small breeding range) with humans simply pushing it over the edge (Chilton 1997), but at that moment one of the connections in the complex web of life on Earth was severed and the web loosened. The more connections that break, the weaker the web becomes. Note that local extinctions (extirpations) also weaken the web. Human laws help prevent species extinctions, but such laws should also protect North America's more than 6000 terrestrial "natural communities" defined by NatureServe, that is, all the "kinds of living nature" defined within the following seven vegetation "classes": forest, woodland, shrubland, dwarf shrubland, herbaceous, nonvascular, and sparse. Protecting some sites in which each natural community occurs protects perhaps 80% of the biodiversity they contain, including species in taxonomic groups that are too little known and too numerous to work on individually (e.g., many invertebrates, lower plants, fungi). This coarse-filter approach to protecting biodiversity must be complemented by the fine-filter approach of protecting individual species. Even though some rocky islets may be protected in a national park by applying the coarse filter, the few used by nesting Labrador Ducks may not be included. This is the key lesson in the Labrador Duck's extinction. NatureServe's network of Conservation Data Centers and Natural Heritage programs excels in such work, but it needs adequate support.

Martin Lipman (Courtesy of Canadian Museum of Nature)

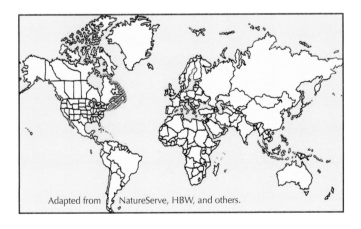

Adapted from NatureServe, HBW, and others.

NatureServe
Conservation Status

Global: .... Extinct
Canada: .... Extirpated
US: .... Extirpated
Mexico: .... Not Applicable

www.natureserve.org

Surf Scoter

Scientific: *Melanitta perspicillata*
Français: Macreuse à front blanc
México: Negreta Nuca Blanca
Order: Anseriformes
Family: Anatidae

This scoter's English name refers to its habit of swimming in or adjacent to the breaking ocean waves, where it "scoots" or dives through the whitecaps. In contrast, the French choose to focus on the white forehead, while the Mexicans are impressed by the white nape.

Surf Scoters have one of the shortest spring pair bonds of the waterfowl. Nesting starts when the birds arrive on their lakes, but males desert their mates after only about three weeks. This doesn't give hens much time to renest if they fail on the first try. Why the males are in such a rush to molt is unknown (Savard et al. 1998), but they probably have a good reason.

When foraging on immobile prey (e.g., mussels), diving waterbirds are thought to complete more of their physiological recovery during the pause on the surface after each dive than do birds foraging on mobile prey (e.g., fish), which may wait until after a series of dives to recover their breath (Beauchamp 1992). The latter might be a useful strategy for making the most of mobile prey while it passes. Surf Scoter individuals in a flock often repeatedly dive and surface simultaneously. This synchronization likely reduces collision risks between diving and resurfacing birds and minimizes predation risks (J-P. Savard, pers. comm.). There may also be social reasons why "skunk heads" flock in the first place (Savard et al. 1998).

Dennis Paulson

Dennis Paulson

Adapted from NatureServe, HBW, and others.

NatureServe
Conservation Status

Global: .... Secure

Canada: .... Secure

US: .... Secure

Mexico: .... Not Applicable

www.natureserve.org

White-winged Scoter

Scientific: *Melanitta fusca*
Français: Macreuse brune
México: Negreta Alas Blancas
Order: Anseriformes
Family: Anatidae

"Large and heavily proportioned, and often riding high in the water, the White-winged Scoter stands out among rafts of scaups on a lake like an aircraft carrier in a fleet of cruisers" (Taylor 2003, 120). It is the largest of North America's three breeding scoter species, its young taking longer to fledge (63–77 days) than any other North American duck species. This may be an adaptation for ducklings to grow in a food-limited environment. Very high nest densities may exist on islands covered with thick vegetation in which females hide their nests. They seldom walk on land and they don't readily flush during approach by an intruder. "On a one-acre island we flushed eight females from their nests amid nettles. The first two had such a slow, laborious takeoff that we realized we should try to pounce upon any others before they got free of the nettles" (Houston 2013). His assistant leaped high to catch one that had flushed, and as his foot landed beside another nest, that scoter flushed and he picked it from the air with his other hand: two scoters caught in a flash! Houston found it to be the easiest duck to catch by hand. Competition between females for nest sites is often high on islands, where they can avoid mammalian predators. Nest sites on islands are often used for six or more years and may even be usurped from other duck species. White-winged Scoters lay eight to 10 eggs, but if intense competition leads two females to lay in the same nest, they will desert it once it contains about 12 eggs (Brown and Fredrickson 1997).

Dennis Paulson

Dennis Paulson

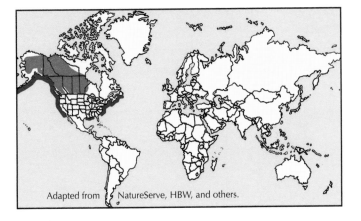

Adapted from NatureServe, HBW, and others.

NatureServe
Conservation Status

Global: .... Secure

Canada: .... Secure

US: .... Secure

Mexico: .... Not Applicable

www.natureserve.org

Black Scoter

Scientific: *Melanitta americana*
Français: Macreuse à bec jaune
México: Negreta Pico Amarillo
Order: Anseriformes
Family: Anatidae

Godfrey's (1986) *Birds of Canada* maps show each species' Canadian breeding range, but the Black Scoter's range was so little known then that he presented no map. He wrote, "widely but locally distributed in summer. Definite breeding records are few" (p.112). Now though, the center of its eastern North American breeding population is known to be located in central Quebec. The Alaskan population was known in Godfrey's time, but is now known to extend east into the northern Yukon and northern Northwest Territories.

All three North American scoter species feed on mollusks, especially in saltwater habitats. In the St. Lawrence River and the Great Lakes, introduced Zebra Mussels (*Dreissena polymorpha*) displace native mussel species, wreak havoc in these ecosystems, and have become an important food source for diving ducks (Hamilton et al. 1994), even influencing staging and wintering distributions of some species (Wormington and Leach 1992). But all that is easy to catch isn't necessarily healthy, for by their filter-feeding behavior mussels accumulate chemicals from the water, which in these waters are many. Selenium, for example, is toxic at high concentrations and it is possible that the recent invasion by Zebra Mussels is attracting migrating and wintering scoters to a hazardous situation (Petrie and Schummer 2002).

Lloyd Spitalnik

The Black Scoter is very vocal. Groups can often be located by the males' sad courtship whistle. Molting sites often have the largest groups, up to several thousand males.

When wing-flapping, the Black Scoter thrusts its head downward, as if its neck is briefly broken, whereas Surf and White-winged Scoters keep the head and bill pointing mostly above horizontal (Bordage and Savard 2011).

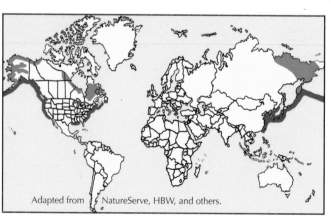

Adapted from NatureServe, HBW, and others.

NatureServe
Conservation Status

Global: .... Secure
Canada: .... Secure
US: .... Secure
Mexico: .... Not Applicable

www.natureserve.org

Long-tailed Duck

Scientific: *Clangula hyemalis*
Français: Harelde kakawi
México: Pato Cola Larga
Order: Anseriformes
Family: Anatidae

In most duck species males molt out of their breeding plumage soon after seeing their mates off to their nests. The resulting eclipse plumage makes the males resemble females for about two months, and in late summer the breeding plumage reappears and is worn from fall into the following spring. Long-tailed Ducks, however, differ from this pattern and have one of the most complex molt sequences found in any bird. Molting is almost continuous from April to October and progresses through various partial molts and interruptions such that four different plumages can be discerned (Sibley 2000).

The light-colored plumage worn from fall to spring makes the Long-tailed Ducks difficult to detect from the air (e.g., from a helicopter). They often dive when an airplane approaches, and can remain submerged longer than any other duck, sometimes reaching depths of more than 60 m (no doubt aided by having the largest heart mass relative to body mass of all waterfowl). Accurate population surveys are further complicated by the species' clumped offshore distribution (Robertson and Savard 2002).

Males use their distinctive call constantly at times; hence the name *clangula*, meaning noise. Thus the Long-tailed Duck and the Common Goldeneye share a Latin word that is used as the genus name for one, but as the species name for the other (Donovan and Ouellet 1993).

John Hoyt

Dennis Paulson

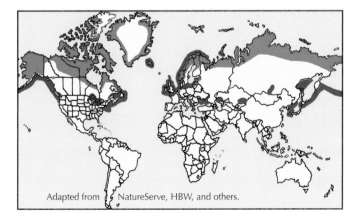
Adapted from NatureServe, HBW, and others.

NatureServe
Conservation Status

Global: .... Secure
Canada: .... Secure
US: .... Secure
Mexico: .... Not Applicable

www.natureserve.org

Bufflehead

Scientific: *Bucephala albeola*
Français: Petit Garrot
México: Pato Monja
Order: Anseriformes
Family: Anatidae

The Bufflehead's apparently fun-loving behavior inspired Bent (1925, 24) to call it the "spirit duck". It is very buoyant, but plunges energetically below the water surface. After a dive it bursts back to the surface like a cork. It explodes into flight from, or even from below, the water surface. When landing, it strikes the water hard and slides to a sudden stop. Bent noted that hunters shot it only as a last resort and often called it "butterball" because it was very fat in autumn.

Its small size may have to do with the Bufflehead's dependence on Northern Flicker cavities, which are too small for other cavity-nesting ducks. Buffleheads use flicker holes in both dead and live trees, but avoid cavities with broken tops. They also avoid larger cavities, thus reducing competition from Common and Barrow's Goldeneyes, which are larger. Buffleheads readily use nest boxes. This facilitates breeding studies by allowing researchers to save time searching for nests and giving them easy access to nest contents (usually via a side door) (Gauthier 1993).

Most duck species feature showy males and drab females, whereas such "sexual dimorphism" is less pronounced or absent in whistling-ducks, geese and, swans. In ducks (e.g., Buffleheads), the showy male must attract a different female each winter and doesn't help raise the chicks, whereas the other waterfowl develop a multi-year pair-bond in which the male helps incubate the eggs and raise the young. Mottled Ducks and American Black Ducks, both close relatives of the sexually dimorphic Mallard, are North American exceptions to this sexual dimorphism "rule".

Steve Zamek

Robert Alvo

Adapted from NatureServe, HBW, and others.

NatureServe
Conservation Status

Global: .... Secure

Canada: .... Secure

US: .... Secure

Mexico: .... Not Applicable

www.natureserve.org

Common Goldeneye

Scientific: *Bucephala clangula*
Français: Garrot à oeil d'or
México: Pato Chillón
Order: Anseriformes
Family: Anatidae

The Common Goldeneye is the only North American duck known to benefit from lake acidification, at least in the short term. Some fish species compete with Common Goldeneyes for invertebrate prey, and when such fish are missing as a result of acid rain, certain invertebrate species thrive and make excellent prey for goldeneyes.

Common Goldeneye breeding biology is well studied, leading to many interesting observations. For example, females in flight have been seen missing their nest cavity openings upon approach and dying from the collision. The smallest clutch known to be incubated was four eggs. Even a 20-egg clutch was incubated, but it involved more than one hen. Single hens never lay more than 12 eggs. The earlier in the season hens start laying, the more eggs they lay, especially in older hens. Hens that lose a nest during egg-laying are unlikely to parasitize (see Redhead account, p. 36) other nests. Clutch size in Common Goldeneyes averages 9.8 eggs, but when parasitized nests are excluded, the average is only 7.1 eggs. Chicks may make peeping calls and tapping sounds from inside the eggs three days before hatching. Ducklings may hatch from eggs cracked by freezing, and eggs may also withstand the hen's absence during incubation for an entire day. Biologists can remove hatching eggs or hatched ducklings from under some hens without displacing them, especially if these are older experienced birds that have previously been handled (Eadie et al. 1995). How much of this observed behavior applies to other, less-known species?

Peter Sproule

The species name *clangula*, meaning noise, refers to the whistling sound of the wings when the bird is in flight. However, males also make a *peent* call during their long December-through-April courtship that strangely resembles the vocalizations of the later-courting American Woodcock and Common Nighthawk. This follows a spectacular backward head-toss and kicking out of water to the side (Eadie et al. 1995).

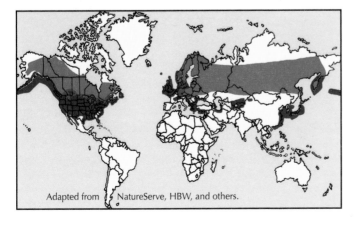

Adapted from NatureServe, HBW, and others.

NatureServe
Conservation Status

Global: .... Secure
Canada: .... Secure
US: .... Secure
Mexico: .... Not Applicable

www.natureserve.org

Barrow's Goldeneye

Scientific: *Bucephala islandica*
Français: Garrot d'Islande
México: Pato Islándico
Order: Anseriformes
Family: Anatidae

This is the most territorial North American duck. The male defends a territory for the breeding pair, but after the magical day of hatching, the female leads her flightless brood over land and water to a different territory that she defends, probably to ensure adequate food and to protect the chicks. Brood territories are often situated on different lakes from pair territories and often a few kilometers away. Females with broods may kill other ducklings of the same or different species, maybe to reduce competition for their chicks' food. A female with chicks is more likely to be aggressive toward strange chicks if the latter differ in age from her own young, although females with chicks younger than 10 days old almost always accept foreign chicks. The Barrow's Goldeneye is the only waterfowl species known to defend winter territories separate from breeding ones. Females that are yearlings or failed nesters will "prospect" for nest sites to be used the next year by visiting various tree cavities, often in groups of two or three birds calling noisily. In central British Columbia, the dominance hierarchy of waterbirds regarding territorial challenging, aggressive posturing, fights, and even murder (beware the spear-like bill of the loon rising from the dark depths) declines from the Common Loon as follows: Red-necked Grebe, Pied-billed Grebe, Horned Grebe, Barrow's Goldeneye, Common Goldeneye, Bufflehead, and American Coot (Eadie et al. 2000).

Larry Master

NatureServe
Conservation Status

Global: .... Secure

Canada: .... Secure

US: .... Secure

Mexico: .... Not Applicable

www.natureserve.org

Adapted from NatureServe, HBW, and others.

Hooded Merganser

Scientific: *Lophodytes cucullatus*
Français: Harle couronné
México: Mergo Cresta Blanca
Order: Anseriformes
Family: Anatidae

"Show me your hood and I'll lay you some eggs", she cooed. All four names of this species above refer to its headdress, which is spectacular in males and attractive nonetheless in females. Most bird species raise their crown feathers at times, but only some have a conspicuous tuft of long feathers that is erect most of the time or that can be raised at will (Terres 1991). These crests are generally considered to function in visual displays. All three of our mergansers, fish-eating diving ducks with a long, thin, serrated, and hooked bill, have crests in some plumage. The male Hooded Merganser uses its crest like a lure during courtship by raising or lowering it to varying degrees and by adjusting the angle at which he shows it to the female. The purpose of the female's crest is unclear. In Common Mergansers, strangely, only the female sports a crest.

Unlike the other two North American mergansers, which feed almost exclusively on fish, the Hooded Merganser has a more diverse diet that includes fish, aquatic insects, and crustaceans, particularly crayfish. It is the smallest of the three species and the only one endemic to North America. Hooded Merganser eggs are unusual in being almost spherical and quite thick-shelled compared to those of the other two mergansers, but the reasons for these differences are unknown. Eggshell thickness is a reliable characteristic for distinguishing these eggs from those of the Wood Duck, another cavity-nester (Dugger et al. 2009).

David Laliberte

Hooded Mergansers are usually seen only in small numbers, but a winter roost at a shallow Florida pond only 45 m in diameter hosted 100–223 birds per night. Much interaction and vocalizing occurred until dark (Barbour and DeGrange 1982).

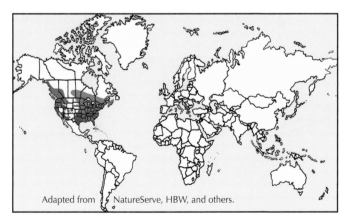

Adapted from NatureServe, HBW, and others.

NatureServe
Conservation Status

Global: .... Secure

Canada: .... Secure

US: .... Secure

Mexico: .... Not Applicable

www.natureserve.org

Common Merganser

Scientific: *Mergus merganser*
Français: Grand Harle
México: Mergo Mayor
Order: Anseriformes
Family: Anatidae

People have been interested in this species largely from fear of competition for tasty salmon and trout, and great efforts have been made in some regions to eradicate it. It often winters as far north as ice-free water allows. Heavier birds and adult males seem to tolerate colder winter temperatures better than others, and thus remain farther north. This is often the first migrant waterfowl species to reach its breeding grounds in the spring, and when females do arrive they are often gregarious—sometimes several hens nest in the same tree. After hatching, broods move gradually downstream over several weeks along rivers or lake systems until reaching wider waters such as larger lakes or estuaries. During the journey that may measure 80 km, broods often merge, probably because of aggression between hens. In the splashing and confusion that occurs, a partial or complete exchange of young among hens may result because of the weak bonds between hens and ducklings and also between chicks and their siblings, especially in larger broods. Amalgamated broods of more than 40 chicks have been seen. Females abandon broods typically 30–50 days after hatching, but sometimes after only one week. The young, which can't fly until 60–75 days old, nevertheless survive well. Large amalgamated broods occasionally contain both Common and Red-breasted Mergansers, and in late fall other diving ducks may join (Alvo 1996; Mallory and Metz 1999). In Europe, "Goosanders" (the Old World name of the Common Merganser) now breed in alpine towns and parks near lake shores in cavities of old trees (Marti and Lammi 1997).

Peter Sproule

Jack Alvo

On their travels downstream, broods often climb down beaver dams. Can you see the well-camouflaged brood?

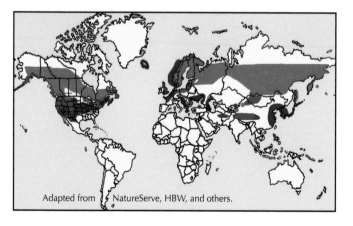

Adapted from NatureServe, HBW, and others.

NatureServe
Conservation Status

Global: .... Secure

Canada: .... Secure

US: .... Secure

Mexico: ..... Possibly Extirpated

www.natureserve.org

Red-breasted Merganser

Scientific: *Mergus serrator*
Français: Harle huppé
México: Mergo Copetón
Order: Anseriformes
Family: Anatidae

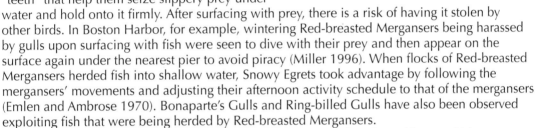

How do two so closely related species as the Red-breasted and Common Merganser with such similar distributions manage to coexist? It may have to do with their opposing breeding phenologies (early in Common vs late in Red-breasted), and/or the Red-breasted's greater use of marine waters. Or consider that in experimental conditions Red-breasteds searched for prey more on the bottom while Commons did so more from the water surface (Titman 1999).

All mergansers have long serrated bills (hence the name *serrator*) with backward-projecting "teeth" that help them seize slippery prey under water and hold onto it firmly. After surfacing with prey, there is a risk of having it stolen by other birds. In Boston Harbor, for example, wintering Red-breasted Mergansers being harassed by gulls upon surfacing with fish were seen to dive with their prey and then appear on the surface again under the nearest pier to avoid piracy (Miller 1996). When flocks of Red-breasted Mergansers herded fish into shallow water, Snowy Egrets took advantage by following the mergansers' movements and adjusting their afternoon activity schedule to that of the mergansers (Emlen and Ambrose 1970). Bonaparte's Gulls and Ring-billed Gulls have also been observed exploiting fish that were being herded by Red-breasted Mergansers.

Bill Schmoker

On 9 August, 1993, a female Red-breasted Merganser with six young was seen on a river on Axel Heiberg Island, Northwest Territories (79°30′N), 900 km north of the species' breeding range. The brood should have had enough time to fledge and start migrating south before freeze-up as long as it first vacated the fresh water for the fjord below (Hofmann et al. 1997). The northernmost record of this species, however, was a lone female on July 2, 1979 near the world's most northerly point of land, in northern Greenland (84°N) (Titman 1999).

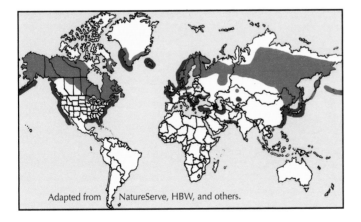

Adapted from NatureServe, HBW, and others.

NatureServe
Conservation Status

Global: .... Secure

Canada: .... Secure

US: .... Secure

Mexico: .... Not Applicable

www.natureserve.org

Masked Duck

Scientific: *Nomonyx dominicus*
Français: Érismature routoutou
México: Pato Enmascarado
Order: Anseriformes
Family: Anatidae

I f you're a biology graduate student looking for a challenge, this may be the species for you. It is one of the most secretive ducks, usually hiding in dense vegetation, rarely taking flight, flying swiftly and low over water and dropping abruptly into vegetation when it does fly, and expertly controlling its buoyancy on water to disappear. Little is known about it. Go catch a few and fit them with radio transmitters so you can record their movements. Then clarify their reproductive biology, habitat requirements, vocal communication, and more. Another useful goal would be to encourage a hunting ban on the Masked Duck, or at least to develop a proper management plan, because only about 3,800 birds are thought to live in the US, breeding only in Texas. Most of its suitable Texan habitat is private, making access difficult, so study it where it is common and more accessible, in Latin America.

The Masked Duck lives in ponds, swamps, marshes, streams, and rice fields with thick vegetation, especially if the surface is covered with leaves. It seems to be mostly vegetarian, often surfacing from below draped, in aquatic plants. Its breeding season is long, with nests found from October until August. It is typical of the "stiff-tailed ducks" like the Ruddy Duck in being small with long pointed tail feathers that have stiff shafts. Unlike the Ruddy Duck, though, it often swims with its tail submerged and avoids open water.

In Texas, alien plants crowd out the native ones that the Masked Duck eats. Although its habitats are routinely cleared of the alien vegetation, native plant communities seldom return afterward. Managers should avoid removing too much vegetation because that also renders the habitat unsuitable for the Masked Duck (Eitniear 1999). The undertaking of a thorough life history study could clearly benefit management of this species throughout its global range.

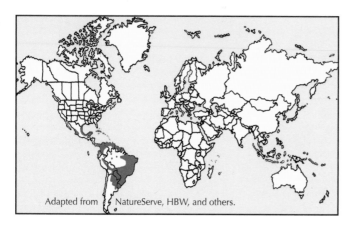

Adapted from NatureServe, HBW, and others.

NatureServe
Conservation Status

Global: .... Secure
Canada: .... Not Applicable
US: .... Vulnerable
Mexico: .... Imperiled

www.natureserve.org

Ruddy Duck

Scientific: *Oxyura jamaicensis*
Français: Érismature rousse
México: Pato Tepalcate
Order: Anseriformes
Family: Anatidae

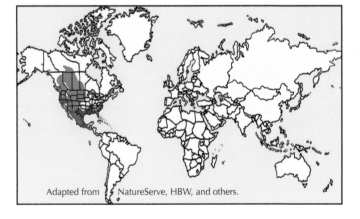

Captive Ruddy Ducks in England escaped in 1952 and spread to mainland Europe where they interbreed with the endangered White-headed Duck (*Oxyura leucocephala*), but programs are in place to try controlling Ruddy Duck numbers there (Hughes 1997).

In North America, however, people love the Ruddy Duck for its charming "bubbling" display, in which it beats its chest with its bill, producing bubbles on the water. Even ducklings do so when startled. Males were thought to participate in brood-rearing, but those that accompany hens with broods seem to be more interested in the hens than in the young.

Compared to other waterfowl, this species has a low ratio of wing area to body size, and its large feet are set far back on the body, both these characteristics aiding in diving. Rather than roll on its side to belly-preen, it stands up and treads water. It is very aggressive, chasing even rabbits on the shore. It has one of the most male-biased sex ratios in waterfowl, and pairs form on the breeding grounds. Hens may use abandoned nests of other species and sometimes lay eggs on top of each other before incubation. Eggs are very large, maybe giving the chicks an advantage when they hatch in the nests of other duck species (see brood parasitism in the Redhead account, p. 36). Once pipping starts, hatching occurs quickly. The highly precocial ducklings are tended by the female for only a short time and she doesn't brood them (Brua 2002).

Edgar T. Jones

Introduced from North America to Great Britain and spreading in western Europe (not shown on map). This male was photographed in France.

Nicole Bouglouan

Adapted from NatureServe, HBW, and others.

NatureServe
Conservation Status

Global: .... Secure

Canada: .... Secure

US: .... Secure

Mexico: .... Secure

www.natureserve.org

Plain Chachalaca

Scientific: *Ortalis vetula*
Français: Ortalide chacamel
México: Chachalaca Oriental
Order: Galliformes
Family: Cracidae

This is our only representative of the Cracidae family, which includes the curassows, guans, and chachalacas. Unlike its Central and South American relatives, it thrives in post-logging brushland. Its range barely extends north into southern Texas where it brings us its family's long, looped trachea responsible for exasperated comments such as artist Christina Lewis's: "It sounded like they were yelling 'Oh shut up! Oh shut up!' and, for the next hour or so, we wished they would shut up!" Males are louder than females, with a longer trachea. Plain Chachalacas call more during the breeding season, during weather changes, after dawn, before dusk, and during full moon nights. They are usually seen in groups of a few to 15 or 20 birds. When not calling they are secretive, running like squirrels along tree limbs through treetops and eating the ripest fruits available, often while upside down. Much of our knowledge is based on birds that have become semi-domesticated, which can be brought about easily through feeding by humans. Birds are domesticated by catching chicks or having chickens incubate eggs. This species may become a nuisance in residential areas and parks with trees, or when it eats crops. It has benefited greatly in Texas from habitat protection aimed at the White-winged Dove, which is hunted and Secure there. Plain Chachalaca populations can tolerate hunting too, despite their Vulnerable US status, if large areas of suitable habitat are protected. In

Juan Bahamon

fact, Peterson (2000) suggests that arrangements be crafted to use funds generated from regulated hunting to conserve chachalacas (see Common Eider account, p. 43, for an example of such an arrangement).

A tree-nester, the Plain Chachalaca often loses nests to strong winds. Once dry, downy chicks cling to branches and climb down to the ground. Bent (1932) thought they cling to their mother's legs, but a later study didn't find this. They can fly short distances a week after hatching. Roosting birds huddle together, all facing the same direction, and often in physical contact with each other (Peterson 2000).

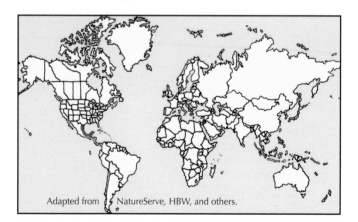

Adapted from NatureServe, HBW, and others.

NatureServe
Conservation Status

Global: .... Secure

Canada: .... Not Applicable

US: .... Vulnerable

Mexico: .... Secure

www.natureserve.org

Mountain Quail

Scientific: *Oreortyx pictus*
Français: Colin des montagnes
México: Codorniz de Montaña
Order: Galliformes
Family: Odontophoridae

Three factors have conspired to prevent us from learning much about this quail: 1) bad luck, 2) its habitat, and, 3) its timidity. It had a rocky scientific initiation with the loss of the type specimen in 1826, and a change in the boundaries of "California" (where early specimens were taken). Also, an important study contained a habitat description of "southern California grasslands", but the species doesn't occur in grasslands. Furthermore, some early key studies were never published and some important early data were lost. The species' original distribution is uncertain because of the lack of data and poor records of introductions. This is a cryptic species that lives in rugged, steep, mountain shrubland. Studying it can somewhat be facilitated if hunting dogs are used or if the population is confined to a small area. Unlike other New World quails, it migrates up to high elevations in the breeding season and back to low elevations in winter to avoid snow cover. It prefers walking or running to flying, even if chased by a ground predator. But if it is surprised by a predator or is about to be caught, it will fly. Reports that it migrates by foot are as yet unproven. Simultaneous double-clutching may be common in Idaho. In that situation, one parent (probably the male) incubates the first clutch while the other incubates the second one. Reproductive success depends greatly on rainfall, with little or no success in dry years. In very dry years this species may not even try to breed. Rain promotes plant growth,

Robert B. Douglas, ©2008 Mendocino Redwood Company

providing food and cover. Mountain Quail habitat can be expanded through the use of "guzzlers", devices that capture precipitation and funnel it into shaded storage tanks that the quails can access. Exit ramps are usually provided to minimize drowning. Up to 40–60 birds, probably from several coveys, have been observed at guzzlers. This is the heaviest and longest quail north of Mexico (no, its long crest isn't included in the standard length of the bird, which is measured from the tip of the bill to the tip of the tail) (Gutiérrez and Delehanty 1999).

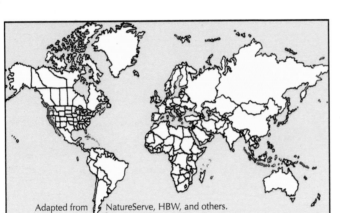

Adapted from NatureServe, HBW, and others.

NatureServe
Conservation Status

Global: .... Secure
Canada: .... Not Applicable
US: .... Secure
Mexico: ..... Apparently Secure

www.natureserve.org

Scaled Quail

Scientific: *Callipepla squamata*
Français: Colin écaillé
México: Codorniz Escamosa
Order: Galliformes
Family: Odontophoridae

All the New World quails have a short, stout, downcurved bill. The family name Odontophoridae, meaning "toothed", refers to the serrations on the lower beak. Like other New World quails the Scaled Quail is gregarious, forming coveys, except when breeding (Sibley 2001). Also called "blue quail" or "cottontop" (for the white cotton-like crest), it is unique-looking in its family with its scaled appearance, the result of black-tipped feathers on the underside, neck, and back. A desert species, it needs land with grasses, shrubs, and forbs. Within this wide array of habitats, it often uses different kinds of plant cover for feeding, loafing, nesting, and night-roosting.

All the New World quail species are hunted to some degree. This, along with grazing in much of their habitat, has stimulated applied research on habitat quality for each species in relation to land use by humans. Overgrazing is the bane of the Scaled Quail's existence because it reduces or eliminates much-needed plant cover. However, some grazing is often acceptable.

Population-boosting techniques that are used with varying degrees of success include: adjusting the timing, duration, and intensity of grazing; changing the livestock species (e.g., cattle, sheep, goat); prescribed habitat burning; reducing shrub density by mechanical means or with herbicides; providing brush piles as surrogates for clumps of shrubs; feeding; providing water; and adjusting the timing and intensity of hunting.

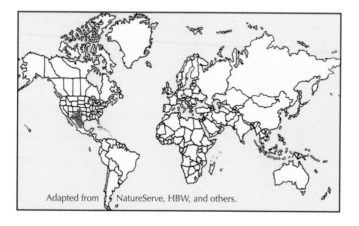

Tom Whetten

In Texas, January and February are the Scaled Quail's "bottleneck" months because reduced cover, bad weather, and low food supplies increase mortality. Hunting at this time becomes additive rather than compensatory. That is, instead of taking birds that would have died of natural causes anyway, such as during the fall when numerous naïve first-year birds are present, many of the birds killed by hunters would otherwise have survived to breed (Dabbert et al. 2009).

Adapted from NatureServe, HBW, and others.

NatureServe
Conservation Status

Global: .... Secure
Canada: .... Not Applicable
US: .... Secure
Mexico: .... Secure

www.natureserve.org

California Quail

Scientific: *Callipepla californica*
Français: Colin de Californie
México: Codorniz Californiana
Order: Galliformes
Family: Odontophoridae

DANY G.

The California Quail's head plume, or topknot, is made up of six forward-curling, comma-shaped black plumes. This is the first thing one may notice when alerted to the species' presence by its characteristic *chi-CA-go* call. The city of Chicago is located quite a bit east of its range. In fact, this quail was originally native to Baja California (Mexico), California, western Nevada, and southern Oregon, but it has been introduced widely elsewhere in western North America and on other continents as a game species. Its small original distribution may have been caused by the unsuitability of other natural North American habitats or by geographic or ecological barriers. One such ecological barrier might have been the presence of the closely related Gambel's Quail, a species that is more tolerant of heat and dry conditions. In places where the California Quail has been successfully introduced, it is usually found on or near agricultural lands.

Attempts to introduce the California Quail into Europe started in 1840, but it wasn't until 1960 that it finally became established in only one European locality, the arable east coast of the Mediterranean island of Corsica off Italy's northwest coast. There it is uncommon and very secretive (Aebischer and Pietri 1997).

Steve Zamek

This species can breed in its first spring, but in some years adults come into reproductive condition two weeks before first-year birds do (Calkins et al. 1999).

Protozoans living symbiotically in the intestines of California Quails, particularly in their ceca (extensions at the junction of the large and small intestine), help the birds to digest roughage and produce essential vitamins such as biotin, riboflavin, niacin, and folic acid. The greater the proportion of roughage in the diet, the larger the ceca. Chicks first obtain these protozoans by pecking at adult feces, and their growth is slowed as a result. This is the price the chicks pay for preparing themselves for an adult diet (Lewin 1963).

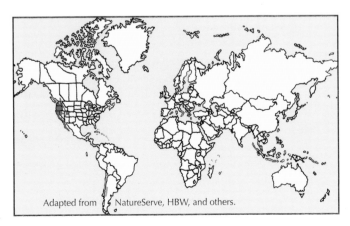

Adapted from NatureServe, HBW, and others.

NatureServe
Conservation Status

Global: .... Secure

Canada: .... Not Applicable

US: .... Secure

Mexico: .... Secure

www.natureserve.org

Gambel's Quail

Scientific: *Callipepla gambelii*
Français: Colin de Gambel
México: Codorniz de Gambel
Order: Galliformes
Family: Odontophoridae

"The Gambel's Quail is also very appropriately called the 'desert quail', for its natural habitat is the hot, dry, desert regions of the southwestern states and a corner of northwestern Mexico" (Bent 1932, 73). "A noteworthy fact in their [life] history is their ability to bear, without apparent inconvenience, great extremes of temperature. They are seemingly at ease among the burning sands of the desert, where, for months the thermometer daily marks 100°F [38°C] and may reach 140°F [60°C] 'in the best shade that could be procured' as Colonel McCall says; and they are equally at home the year round among the mountains, where snow lies on the ground in winter" (Coues 1874, 434). "In inhabited regions, in places where cattle trails lead to water, the Gambel's Quail's pretty footprints call up pleasant pictures of morning processions of thirsty little 'black-helmeted' pedestrians talking cheerfully as they go" (Bailey 1928). Further study is needed to determine how this species is able to live in such harsh conditions.

Gambel's Quails were easily trapped by First Nations people and European settlers until the early 1900s when the practice was made illegal in the US. They are still trapped legally in Sonora, Mexico, though perhaps to a lesser degree than in the past. They do well and reproduce in Sun City, Arizona, and in other residential areas where domestic cats (*Felis catus*) and dogs (*Canis lupus familiaris*) are kept indoors, but they are absent from similar areas where pets roam outdoors (Gee et al. 2013). Domestic cats kill 1.4–3.7 billion birds (all species combined)

Richard Higgins

and 6.9–20.7 billion mammals annually in the US alone, the majority being taken by feral cats as opposed to human-owned ones. Cats are likely the greatest source of human-caused mortality for US birds and mammals (Loss et al. 2013). Ground-nesting birds such as quail are more susceptible to predation than are other birds.

Expansion of exotic grasses in the Sonoran Desert has led to more frequent wildfires, and it would be useful to study the effects of this change on Gambel's Quail populations (Gee et al. 2013).

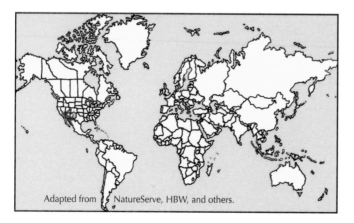

Adapted from NatureServe, HBW, and others.

NatureServe
Conservation Status

Global: .... Secure

Canada: .... Not Applicable

US: .... Secure

Mexico: .... Secure

www.natureserve.org

Northern Bobwhite

Scientific: *Colinus virginianus*
Français: Colin de Virginie
México: Codorniz Cotuí
Order: Galliformes
Family: Odontophoridae

ON 3 WE EXPLODE, GOT IT?

Northern Bobwhite coveys form at the end of the breeding season, with first-year birds and adults present together in coveys in the "fall shuffle". Individuals may switch coveys, and the degree to which covey members are related isn't known (Brennan 1999). "When a flock of quails suddenly bursts into the air from almost underfoot the effect is startling and gives the impression of great strength and speed. They have been referred to as feathered bombshells. Such sudden flights of a whole bevy in unison are due to the fact that they have crouched, trusting to their wonderful powers of concealment, until the very last moment when they are forced to make a quick get-away" (Bent 1932, 20).

In this species' "ambisexual polygamy" breeding system, both sexes incubate and raise broods with more than one mate during the breeding season, but only a limited proportion of the population's females (about 40%) and males (about 20%) are polygamous.

One of the most intensively studied birds in the world, this very important game bird has been tested for the physiological and behavioral effects of pesticides on wildlife. The Northern Bobwhite was the subject of the first modern systematic study of a wild animal's life history in relation to environmental and habitat factors that influence its abundance. It was also used to study the usefulness of prescribed fire for habitat management. Even its color preferences for food and grit were studied: captive females avoided dyed food items whereas males didn't, and all the birds seemed to prefer yellow grit over green grit to help with digestion (Brennan 1999).

Edgar T. Jones

Emily Pipher

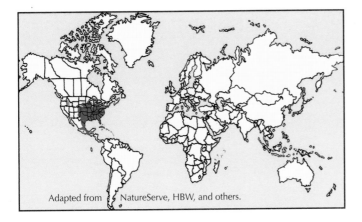

Adapted from NatureServe, HBW, and others.

NatureServe
Conservation Status

Global: .... Secure
Canada: .... Critically Imperiled
US: .... Secure
Mexico: .... Secure

www.natureserve.org

Montezuma Quail

Scientific: *Cyrtonyx montezumae*
Français: Colin arlequin
México: Codorniz Moctezuma
Order: Galliformes
Family: Odontophoridae

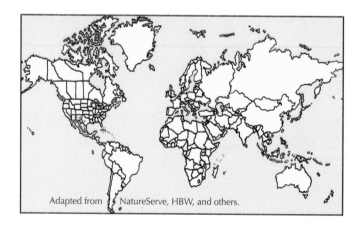

The genus name *Cyrtonyx* is from the Greek *kyrtos*, meaning arched, and *onyx*, meaning claw, referring to the long, curved claws that allow Montezuma Quails to move with agility over rough, rocky terrain. You can hike in excellent habitat for days without encountering this species because even the contrastingly colored males are well camouflaged. It is therefore not a well known or adequately studied species, and a technique to find and observe it in the wild is needed.

It feeds only on the ground, usually digging for food with its feet by using its claws to scratch and dig holes. Thus it leaves evidence of its work in the form of cone-shaped diggings 1–3 cm deep, with one side of the cone excavated as a trench that exposes plant bulbs and roots. It also leaves slightly larger holes and fan-shaped piles of tailings. What you may see while investigating are several square meters of these diggings, which offer some clues about its life history. Adults scratch up seeds, bulbs, and

Larry Master

insects for the young until the latter are about two weeks old. Adults peck at the ground near chicks, often moving food items, but without placing them into the chicks' bills. Rather, the young peck at the adult's bill and take the food items themselves.

Montezuma Quails can jump explosively straight up. Traps for this species have to be modified to include a soft, elastic netting on the ceilings to prevent birds from injuring their heads when they do so. For captive birds, enclosures must be tall enough to allow for these jumps, and the birds' wings have to be clipped to prevent individuals from harming themselves when they try flying inside the small enclosure (Stromberg 2000).

Adapted from NatureServe, HBW, and others.

NatureServe
Conservation Status

Global: ...... Apparently Secure
Canada: .... Not Applicable
US: .... Apparently Secure
Mexico: .... Vulnerable

www.natureserve.org

Ruffed Grouse

Scientific: ***Bonasa umbellus***
Français: Gélinotte huppée
México: Gallo de Collarín
Order: Galliformes
Family: Phasianidae

"During the first warm days of early spring the wanderer in our New England woods is gladdened and thrilled by one of the sweetest sounds of that delightful season, the throbbing heart as it were, of awakening spring. On the soft, warm, still air there comes to his ears the sound of distant, muffled drumming, slow and deliberate at first, but accelerating gradually until it ends in a prolonged, rolling hum" (Bent 1932, 141–142). The male performs his unique drumming display from a slightly raised platform by bracing his tail, spreading his wings, and rotating them forward, then quickly backward, up to 50 times during the 8–11 second rendition, each cycle of "the motor" creating a compression, then a release of air pressure.

The northern part of the Ruffed Grouse's range overlaps that of the Snowshoe Hare (*Lepus americana*), which is well-known for its 10-year population cycles caused by interactions between hares and their food and predators (Boutin et al. 1986). Hare numbers increase in response to a large food supply. This is followed by an increase in hare predator populations, namely Northern Goshawks and Great Horned Owls. Corresponding increases in these two predators' numbers cause a decline in hares, at which point the predators turn to Ruffed Grouse for food, which causes grouse numbers to decline. Thus the grouse are caught up in this 10-year population cycle. Closing the hunting season doesn't interrupt the 10-year cycle. Of note, in the southern part of the Ruffed Grouse's range, Snowshoe Hares are absent and grouse populations don't follow this pattern (Rusch et al. 2000).

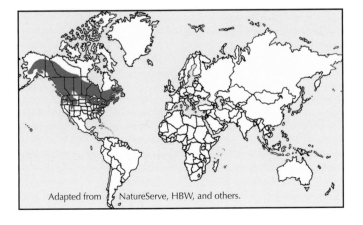

Don Wigle

Larry Master

Adapted from NatureServe, HBW, and others.

NatureServe
Conservation Status

Global: .... Secure

Canada: .... Secure

US: .... Secure

Mexico: .... Not Applicable

www.natureserve.org

Greater Sage-Grouse

Scientific: ***Centrocercus urophasianus***
Français: Tétras des armoises
México: Gallo de Artemisas Común
Order: Galliformes
Family: Phasianidae

"D.E. Brown tells me of a nest found by a sheep herder. The bird did not flush from the nest until the sheep were all around her; she then flushed with a great noise, scattering the sheep in all directions. This habit may often prove very useful in preventing cattle from trampling on the eggs" (Bent 1932, 303).

Leks are traditional courtship areas where many males of a species try to attract females for mating by performing courtship displays from small sites that each male defends. In most species only a few males earn most of the copulations. Lekking behavior occurs not only in birds but also in insects, amphibians, and mammals. In this mating system the only breeding role of males is to fertilize females. Greater Sage-Grouse lek sites always have less vegetation than surrounding habitats and now often occur in or near cultivated fields, gravel pits, or roads.

In this species the average peak number of males per lek is 14–70, but peak counts can be in the hundreds of males. The average distance between neighboring leks is 1.4–5.1 km. Yearling males can attend leks, but older males are much more successful at breeding. Male numbers at leks increase over the breeding season as yearlings join. Copulation lasts only a few seconds, and males may mate with other females within two minutes, sometimes mating more than 20 times in a morning. One male copulated 169 times in one season. Many males may loiter around successful males in hopes of a "spillover effect". The female lays her first egg 3–14 days after copulation (Schroeder et al. 1999).

Todd Black | Todd Black | Emily Pipher

Adapted from NatureServe, HBW, and others.

NatureServe
Conservation Status

Global: .... Vulnerable
Canada: ....Critically Imperiled
US: .... Vulnerable
Mexico: .... Not Applicable

www.natureserve.org

Gunnison Sage-Grouse

Scientific: *Centrocercus minimus*
Français: Tétras du Gunnison
México: Gallo de Artemisas de Gunnison
Order: Galliformes
Family: Phasianidae

The Greater Sage-Grouse and Gunnison Sage-Grouse live only in sagebrush eco-systems. Sagebrush is a common name given to various shrubs and herbaceous plants of the genus *Artemesia*. The two grouse taxa were split into two distinct species by the American Ornithologists' Union (2000) based on genetic differences, the smaller size of the Gunnison, differences in male courtship behavior, the more elaborate head plumes of the male Gunnison (recurved crest instead of short, scattered, erect filoplumes), and the broader white banding on the Gunnison's tail feathers.

Whereas the Greater Sage-Grouse is ranked Vulnerable at the global scale, the Gunnison is ranked Critically Imperiled. A number of factors are considered when determining the conservation status rank (see Introduction, pp. 9–10), but in this case the main reasons for this difference are: the Gunnison's much smaller distribution (4800 vs 670,000 km²); its much smaller number of populations (eight vs "more than 80"); its much smaller number of populations with good viability and/or integrity ("perhaps one" vs 13–125); and its much smaller number of individuals (4400 vs 536,000) (NatureServe 2014b). Ranks already assigned at other geographic scales can also aid in ranking because the criteria are identical at the three scales. Thus, once the Gunnison is ranked globally Critically Imperiled (G1), it shouldn't be ranked nationally Imperiled (N2) or Vulnerable (N3). Rather, for any country in which it breeds, it should be ranked Critically Imperiled (N1). Likewise, for any state in which it occurs, it should be ranked Critically Imperiled (S1). Indeed, it is ranked G1N1S1 in Colorado. Its Utah rank of G1N1S2, however, indicates disagreement between the state rank and the other two ranks. NatureServe's conservation status ranks and their underlying data have many applications, including environmental impact assessments, the designation of species under species-at-risk legislation, and natural site evaluation for conservation priority.

The best way to see and learn about the Gunnison Sage-Grouse is to contact: www.siskadee.org.

Noppadol Paothong

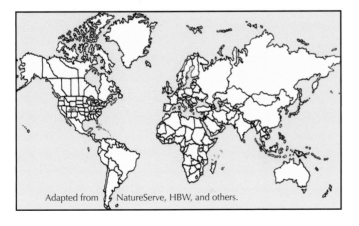

Adapted from NatureServe, HBW, and others.

NatureServe
Conservation Status

Global: ..... Critically Imperiled

Canada: .... Not Applicable

US: .... Critically Imperiled

Mexico: .... Not Applicable

www.natureserve.org

Spruce Grouse

Scientific: *Falcipennis canadensis*
Français: Tétras du Canada
México: Gallo Canadiense
Order: Galliformes
Family: Phasianidae

The Spruce Grouse "remains so woefully ignorant of the destructive nature of the human animal that, unlike its cousin the Ruffed Grouse, it rarely learns to run or fly away but allows itself to be shot, clubbed, or noosed, and in consequence has earned for itself the proud title of 'fool hen'. As a result wherever man appears the Spruce Grouse rapidly diminishes in numbers and in the vicinity of villages or outlying posts is not to be found" (Townsend, C. W., in Bent 1932, 120–121). How has this bird survived over evolutionary time given its seemingly weak sense of personal security? Its cryptic coloration, particularly in females, combined with its tendency to remain motionless until closely approached, must be an effective strategy for avoiding its natural predators, which presumably didn't include humans.

The Spruce Grouse lives in conifer forests that are regenerating after fire. Being mostly a herbivore, its main food is pine needles, and in winter this may be its only food. Where pines aren't readily available it eats spruce needles. It tends to forage in the mid-crown of trees, perhaps because needles there are more nutritious. These branches also provide sturdy support, and the bird can watch for avian predators while remaining concealed. When eating larch needles in autumn, it prefers yellowing ones over green ones. Its preferences may be linked to nutrient content, the presence of certain compounds, or the ease of browsing. The Spuce Grouse seems to apply the same preferences when selecting among individual trees of the same species, but some trees adjacent to heavily browsed ones are mysteriously ignored. Individual grouse differ in their

Giff Beaton

degree of selectivity. Twigs are gripped between mandible tips and broken off with a flick of the head. This wears off the upper mandible tip by the end of winter, resulting in variance in culmen length measured over the year.

Only hens incubate. They travel an average of 83 m from the nest to forage and defecate during 1–6 recesses taken per day. Recesses take an average of 26 minutes and their length is correlated with egg cooling. Egg temperature declines an average of 5.5°C per recess (Boag and Schroeder 1992).

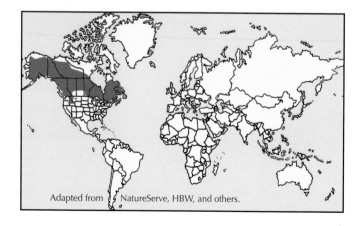

Adapted from NatureServe, HBW, and others.

NatureServe
Conservation Status

Global: .... Secure
Canada: .... Secure
US: .... Secure
Mexico: .... Not Applicable

www.natureserve.org

Willow Ptarmigan

Scientific: *Lagopus lagopus*
Français: Lagopède des saules
México: Lagópodo Común
Order: Galliformes
Family: Phasianidae

The genus name of the world's three ptarmigans, *Lagopus*, means hare's foot. In winter, ptarmigans' feet are thickly covered with long, hair-like feathers, they have a fringe of scales along each toe, and the claws are large, all this additional surface area acting like snowshoes by enabling the birds to walk on soft snow without falling through. Ptarmigans molt at least three times per year, which gives them cryptic coloration in all seasons (e.g., white in winter).

In winter, the Willow Ptarmigan spends up to 80% of its time in snow burrows to hide and keep warm. "And nature, as if realizing the perils of the ptarmigan asleep, has taught it to plunge beneath the cold drifts to escape the cold, and to *fly at*, not *walk to*, the chosen drift so that there will be no telltale trail for some keen nose to follow to the sleeping-place. And this the bird invariably does, going at speed and butting its way into the snow, leaving never a print to betray its retreat from which it *flies* forth in the morning" (Sandys and Van Dyke 1904, 224). Is it possible that this practice of flying into snowbanks is learned through play? In young animals including birds, play enhances learning skills and other behaviors in a safe situation in preparation for adulthood. In Willow Ptarmigans, play can include crouching with the head low and extended forward with the eye combs partly erect, tilting the bill up and wagging the head, jumping around erratically, flapping wings, or jerking one wing then the other.

John Hoyt

The breeding range of the Willow Ptarmigan overlaps that of the Rock Ptarmigan in both the New World and the Old World, but the two species are usually separated by habitat—the Willow in wetter, shrubbier habitat, the Rock in drier, rockier areas. When the two species meet, males may interact with aggression, but Willow Ptarmigans dominate over Rock Ptarmigans.

As in other ptarmigans, though unusual in birds, the Willow Ptarmingan lays eggs wet with pigment in which female feather prints may be seen. As the red pigment dries it turns blackish brown and persists on discarded shells until at least the following year (Hannon et al. 1998).

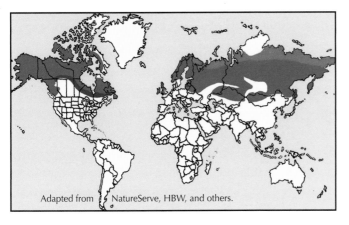
Adapted from NatureServe, HBW, and others.

NatureServe
Conservation Status

Global: .... Secure
Canada: .... Secure
US: .... Secure
Mexico: .... Not Applicable

www.natureserve.org

Rock Ptarmigan

Scientific: *Lagopus muta*
Français: Lagopède alpin
México: Lagópodo Alpino
Order: Galliformes
Family: Phasianidae

This is one of only a few bird species living year-round on the Arctic tundra. It copes very well in the harsh, windswept conditions. Unlike its close relatives, it has a layer of fat in winter that provides energy when food is scarce or difficult to access. It avoids deep snow by choosing ridges and slopes that are exposed to wind, and by foraging in areas where Caribou (*Rangifer tarandus*), Muskox (*Ovibos moschatus*), or hares have exposed the vegetation. According to some Inuit hunters, it may even follow Caribou for this purpose. In late autumn and winter it descends to lower elevations for conditions that are more favorable. (In late summer, however, it often moves to higher elevations, probably to take advantage of food that has recently become exposed by retreating snow.) Females on nests during spring blizzards may remain there continuously for 36 hours. When an incubating female is approached, her pulse rate and breathing slow down, which makes her more difficult for the predator to hear and smell. When the predator gets too close, these vital signs rise suddenly as she considers flying. Often the critical distance is only 1–2 m, and some females don't move until touched. Because females molt out of their white winter plumage before males, incubating females are cryptic at the same time when males are white. Territorial males may be visible from 1–2 km away, whereas females sitting on exposed nests are often difficult to find from less than 2 m away, and Arctic Foxes (*Vulpes lagopus*) may walk right past them. The males' breeding plumage remains white into the females' incubation period even though males dust-bathe, but the white becomes soiled rapidly when females are no longer sexually receptive. Soiled males look and act much less conspicuous than do clean ones; thus soiling may help to conceal them from predators when the clean white plumage is no longer needed for territoriality and pair bonding (Montgomery and Holder 2008).

Darroch Whitaker

"Sometimes I found my old footprints taken possession of by the ptarmigans as night-quarters" (Manniche 1910). This would be one of those rare occasions when the human "footprint" has some benefit for the natural world.

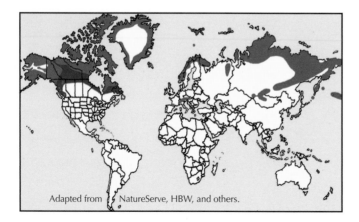

Adapted from NatureServe, HBW, and others.

NatureServe
Conservation Status

Global: .... Secure

Canada: .... Secure

US: .... Secure

Mexico: .... Not Applicable

www.natureserve.org

White-tailed Ptarmigan

Scientific: *Lagopus leucura*
Français: Lagopède à queue blanche
México: Lagópodo Cola Blanca
Order: Galliformes
Family: Phasianidae

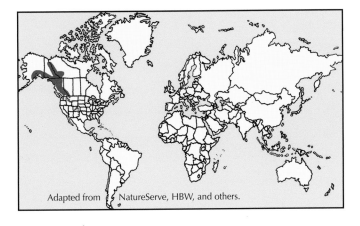

This ptarmigan is an alpine resident in habitats occurring above the tree line, though it makes vertical movements from high elevation breeding sites to lower elevation, albeit treeless, wintering areas. It tolerates a wide range of temperatures without using up extra energy to control its body temperature.

The White-tailed Ptarmingan usually walks rather than flies. When foraging, it walks slowly and creeps low and cautiously. In very cold weather it chooses places that are a few degrees warmer than the ambient temperature. Part of its strategy for living in these cold, windy conditions where the air is thin is simply to avoid exerting itself. We humans understand well, for when we visit the Rocky Mountains at 3,000 m altitude we can experience some real trouble breathing while hiking. Indeed, the White-tailed Ptarmigan lives a rather sedentary lifestyle. However, it may find the energy to dash across areas that are exposed to aerial predators, and it can hop onto rocks with the aid of its flapping wings (Braun et al.1993).

As in the other two ptarmigans, its North American distribution is thought to have changed little because ptarmigans live in places that are largely undeveloped. Occurring farther south than the other two ptarmigans, however, the White-tailed Ptarmigan comes into more contact with humans because alpine regions may have developments such as roads, mines, water reservoirs, snow catchment fences, microwave relay stations, off-road vehicles, ski areas, and even some grazing livestock. Management practices that can help this species include adjusting hunting seasons in favor of seasons when alpine areas are less accessible, rerouting roads and ski trails, relocating snow catchment fences, or minimizing livestock grazing effects by reducing grazing or changing its timing (Braun et al. 1993).

A note for birders: this species is sometimes seen below the tree line (Sinclair et al. 2003; Schroeder 2005).

Frances Alvo

NatureServe
Conservation Status

Global: .... Secure
Canada: .... Secure
US: .... Secure
Mexico: .... Not Applicable

www.natureserve.org

Adapted from NatureServe, HBW, and others.

Dusky Grouse

Scientific: *Dendragapus obscurus*
Français: Tétras sombre
México: Gallo de las Rocosas Oriental
Order: Galliformes
Family: Phasianidae

"Blue Grouse" occur in the mountains of western North America. Historically the two groups that their eight subspecies fall into have been variously treated as one or two species. Now they are treated by the American Ornithologists' Union as two species based on differences in behavior, morphology, vocalizations, and genetics (Banks et al. 2006). Dusky Grouse males usually call from the ground, whereas Sooty Grouse males usually call from trees. In the Dusky the inflatable neck sacks are smoothly textured, reddish purple, and surrounded by a wide white-feathered border, but they are roughly textured, yellow, and surrounded by a narrow white-feathered border in the Sooty. Dusky males have a five-syllabled, soft, low-pitched vocalization that carries only 30–40 m, whereas Sooty males have a six-syllabled loud high-pitched vocalization that can be heard up to 500 m away. Other differences include 20 tail feathers in the Dusky vs 18 in the Sooty, and grayish chicks vs yellowish ones. The Dusky lives in the Rocky Mountains, whereas the slightly darker Sooty lives in the coastal ranges of the west coast of North America. There is little range overlap (Sibley 2000; Zwickel and Bendell 2005).

As in other species pairs that have been split recently, much of the existing literature is presented for the two species combined and it is often difficult to know to what degree the information applies to one species or the other.

Gale (1892, 42) wrote of the Dusky Grouse that in the mating season the male "takes his stand upon a rock, stump, or log and … performs his growling or groaning, I don't know which to call it, having the strange peculiarity of seeming quite distant when quite near, and near when distant; in fact, appearing to come from every direction but the true one". Surprisingly little has been published on ventriloquism in birds, but apparently some birds use this ability to warn other birds of predators while remaining hidden themselves.

Hans Westerlaken

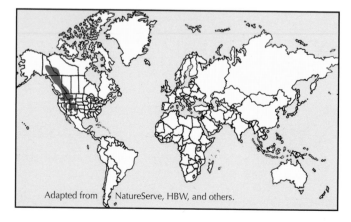

Adapted from NatureServe, HBW, and others.

NatureServe
Conservation Status

Global: .... Secure

Canada: .... Secure

US: .... Secure

Mexico: .... Not Applicable

www.natureserve.org

Sooty Grouse

Scientific: *Dendragapus fuliginosus*
Français: Tétras fuligineux
México: Gallo de las Rocosas Occidental
Order: Galliformes
Family: Phasianidae

In contrast with most other birds that make vertical migrations, the Sooty Grouse moves in autumn from relatively open breeding areas in the lowlands to more dense conifer forest in the mountains. When migrating, it walks and flies (Bent 1932). Anthony's (1903, 27) following description of the fall migration of Dusky Grouse perhaps equally applies to the Sooty Grouse. "Before the middle of August, the migration is in full swing and flocks are seen each evening passing over Sparta [Oregon]. Frequently they alight in the streets and on the housetops. I recall with a smile the memory of a flock of a dozen or more which lit one evening in front of the hotel. For a time pistol bullets and bird shot made an accident policy in some safe company a thing to be desired, but strange to relate none of the regular residents of the town were injured. The same may be said of most of the grouse, though one, in the confusion, ran into the livery stable and took refuge in a stall, where it was killed with a stick". As for the time after migration, "It is a remarkable bird with habits all its own. Throughout the winter months the birds are seldom seen, even in places where they are most abundant, for at this season they retire to the heavy fir timber and spend their time very quietly high up in the trees. During this season they feed almost exclusively on fir buds and do not even descend to the ground to drink, for the abundant rainfall makes it possible for them to quench their thirst in the treetops. Personally, I do not think that the grouse ever voluntarily comes to the ground during these months of retirement. Only when by accident they are disturbed, as when woodsmen are felling trees, are they likely to be seen. In sections where the grouse are very abundant you may pass through the woods day after day, and unless you understand their ways, never suspect that such a bird is present" (Haskin, L. L., in Bent 1932, 113).

Christian Artuso

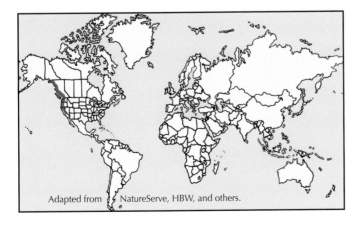

Adapted from NatureServe, HBW, and others.

NatureServe
Conservation Status

Global: .... Secure
Canada: .... Secure
US: .... Secure
Mexico: .... Not Applicable

www.natureserve.org

Sharp-tailed Grouse

Scientific: *Tympanuchus phasianellus*
Français: Tétras à queue fine
México: Gallo de las Praderas Coludo
Order: Galliformes
Family: Phasianidae

A humorous and accurate description of this species' elaborate courtship behavior, which resembles that of its closely related species, is given by Bent (1932, 291–292). "These birds have favorite spots, generally small knolls, to which they resort for this purpose every spring [30–45 minutes before sunrise in fall and spring for two to four hours]; these are known as 'dancing hills'" [now called leks]. Elliott (1897, 129–130) continues: "The birds, both males and females,… go through a performance as curious as it is eccentric. The males, with ruffled feathers, spread tails, expanded air sacs on the neck, heads drawn toward the back, and drooping wings (in fact, the whole body puffed out as nearly as possible into the shape of a ball on two stunted supports), strut about in circles, not all going the same way but passing and crossing each other in various angles.… Suddenly they become quiet and walk about like creatures whose sanity is unquestioned, when some male again becomes possessed and starts off on a rampage, and the 'attack' from which he suffers becomes infectious and all the other birds at once give evidences of having taken the same disease". "The whole performance reminds one so strongly of a Cree dance as to suggest the possibility of its being the prototype" (Seton 1890).

Edgar T. Jones

Males new to a lek establish territories as such: they occupy an area near the lek edge for three to five days, then walk onto the lek and sit or crouch at a spot for one or two days; over the next day or two they display to, and defend a small territory against, more peripheral males; finally they display and defend the territory against all adjacent males. The latter disrupt 26–50% of copulation attempts by pushing the male off the female. Males that occupy central territories often have larger sperm volumes, more motile sperm, and higher mating success than do peripheral ones (Connelly et al. 1998).

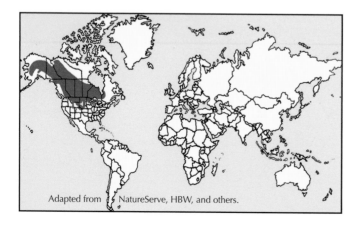

Adapted from NatureServe, HBW, and others.

NatureServe
Conservation Status

Global: .... Secure

Canada: .... Secure

US: .... Apparently Secure

Mexico: .... Not Applicable

www.natureserve.org

Greater Prairie-Chicken

Scientific: *Tympanuchus cupido*
Français: Tétras des prairies
México: Gallo de las Praderas Mayor
Order: Galliformes
Family: Phasianidae

"A prairie cock when in the lists [in the mood] is a strikingly conspicuous creature; he wears no adornment which cannot be concealed at a moment's notice. The sight of a passing hawk changes the grotesque, beplumed, be-oranged bird into an almost invisible squatting brownish lump, so quickly can the feathers be dropped and air sack deflated" (Chapman 1908a).

The story of one of the Greater Prairie-Chicken's geographically isolated subspecies, the Heath Hen (*Tympanuchus cupido cupido*), is one of the earliest North American conservation stories, for legislation was passed in 1791 to protect it from market hunting. It lived on the Atlantic coastal plain from Massachusetts south to Virginia, but declined rapidly after European settlement. "It is the first time in the history of ornithology that a bird has been studied in its normal environment down to the very last individual" (Gross, A. O., in Bent 1932, 269). After December 8, 1928, apparently only one bird, a male, survived. It was last seen on its lek on the island of Martha's Vineyard, off Massachusetts, on March 11, 1932. This being early in the breeding season, the failure of his many faithful observers to see him there again suggested that he died soon afterwards. Another isolated subspecies, the Attwater's Prairie-Chicken (*T. c. attwateri*), has been reduced to three small, isolated populations in southeastern Texas. After the market hunting era, the main cause of decline of the entire species, now composed mainly of the third subspecies, *T. c. pinnatus*, was the conversion of its habitats for crops (Johnson et al. 2011).

Rich Phalin

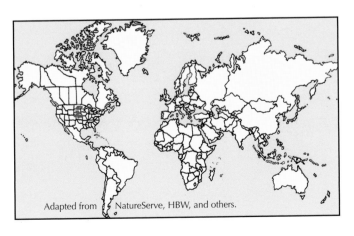

Adapted from NatureServe, HBW, and others.

NatureServe
Conservation Status

Global: ...... Apparently Secure

Canada: .... Extirpated

US: .... Apparently Secure

Mexico: .... Not Applicable

www.natureserve.org

Lesser Prairie-Chicken

Scientific: *Tympanuchus pallidicinctus*
Français: Tétras pâle
México: Gallo de las Praderas Menor
Order: Galliformes
Family: Phasianidae

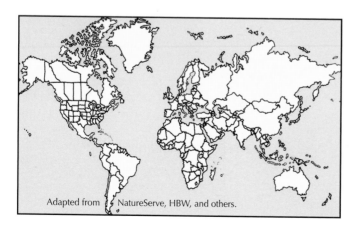

The Lesser Prairie-Chicken has recently been drawing considerable attention and is now listed as "Threatened" under the US Endangered Species Act. This could potentially have great financial implications if, for example, the development of wind energy were to be curbed in an effort to help the species. At the time of writing, the potential effects of wind energy infrastructure on this species continue to be studied and debated.

Like its close relatives, the Lesser Prairie-Chicken is a lekking species in which a small number of dominant males win most of the copulations with females. Some leks are known to have been used for at least 30–40 years. Dominant males occupy small territories in the center of the lek, while subordinate ones have larger territories on its periphery. Features of the vegetation and substrate help males delineate territorial boundaries; thus when the soil is sandy and unstable the boundaries often change accordingly. Individual male territories may be as small as 3.5 m in diameter, and some males have been documented using exactly the same territory in the same lek for two years. Males attend leks from January to June and from September to November, usually defending the same territory in both seasons; females attend less often in the fall. Males at times display to females off leks, and some mating

Texas Parks and Wildlife Department

may occur there. The number of males at a lek peaks from sunrise until 105 minutes after sunrise. In spring, males and females attend leks and display also in the evening, with the number of males reaching a maximum less than one hour before sunset. Females often visit leks in small flocks. Once they arrive, a dominance hierarchy may develop with dominant hens chasing other hens away from dominant males. Interestingly, females often nest closer to a lek other than the one at which they recently mated (Hagen and Giesen 2005).

Adapted from NatureServe, HBW, and others.

NatureServe
Conservation Status

Global: .... Vulnerable
Canada: .... Not Applicable
US: .... Vulnerable
Mexico: .... Not Applicable

www.natureserve.org

Wild Turkey

Scientific: *Meleagris gallopavo*
Français: Dindon sauvage
México: Guajolote Norteño
Order: Galliformes
Family: Phasianidae

The Wild Turkey is North America's largest upland game bird and the only New World bird to be domesticated worldwide, having been first brought to Europe in the 1500s. It is so commonly known that it seems to have its own glossary that includes terms such as: tom, snood, beard, dewlap, and poult. In 1989 alone, 260 million domestic turkeys were gobbled up in the US, grossing US$2.24 billion. One reason why domesticated turkeys cannot successfully be reintroduced to the wild may be that chicks lose the habit of keeping absolutely still at the hen's command. Instead of domestication, the practice that has led to one of the greatest species comebacks in North America is that of trapping wild individuals in one place and reintroducing them elsewhere. In Ontario, for example, almost 300 turkeys were obtained from various US states in the 1980s by trading other wild animals for them. Small numbers of these turkeys were reintroduced into a number of areas in southern Ontario, where the population has grown to about 80,000 birds—not bad for a species that was on the brink of extinction in 1900 (Eaton 1992). "A common method of capture … was to trap them in a … pen made of logs…. A trench was dug sloping gradually down under the log wall and up into the pen. Corn … was sprinkled along this trench and plenty of it spread on the inside of the pen to tempt the turkeys to enter. When, after eating all they wanted, they attempted to escape and constantly looked upward for an opening, but seldom, if ever, had sense enough to crawl out the way they had come in" (Bent 1932, 337). What "turkeys"!

This species has a very good fossil record because it was eaten by First Nations people and settlers, and also because it is very large and its bones are dense (Eaton 1992).

Diane Lepage

Frank Phelan

Four birds walked from top left to bottom right. The long lines were made by toe drag.

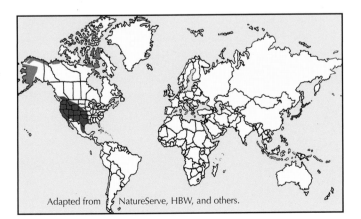

Adapted from NatureServe, HBW, and others.

NatureServe
Conservation Status

Global: .... Secure

Canada: .... Secure

US: .... Secure

Mexico: .... Vulnerable

www.natureserve.org

Red-throated Loon

Scientific: *Gavia stellata*
Français: Plongeon catmarin
México: Colimbo Menor
Order: Gaviiformes
Family: Gaviidae

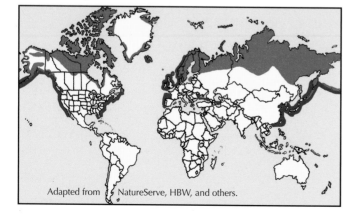

In all five loon species, the legs are attached very far back (posteriorly) on the body. Loons are thus awkward on land. They are so aquatic that they might prefer to nest on or in the water, but they can't, so they always nest right next to it to be able to easily come and go. Even though they winter in salt water and breed in fresh water, they can't nest on marine shorelines or along tidal fresh water because the nest would become flooded twice per day.

The aberrant Red-throated Loon, the smallest loon, takes off easily, whereas Common Loons patter along the lake surface to take off, then circle the lake or bay several times to clear the trees. Red-throated Loons breed on small tundra and Arctic ponds that are usually fishless, and these runways are short. They often make regular flights to and from these ponds, ferrying food in for the young from the sea, lakes, or rivers, one fish at a time. Red-throated Loons don't brood young chicks on their backs as do other loons (Barr et al. 2000), perhaps because few chick predators such as large fish and turtles lurk in these ponds.

When I was conceiving this book in 2001, this species was to appear first. In July 2003, the American Ornithologists' Union moved the waterfowl and chicken-like birds to the front of the evolutionary order based on new data (Banks et al. 2003). The Black-bellied Whistling-Duck usurped first place. Such decisions, though justified, often meet resistance at first by birders.

Edgar T. Jones

Dennis Paulson

Adapted from NatureServe, HBW, and others.

NatureServe
Conservation Status

Global: .... Secure

Canada: .... Secure

US: .... Secure

Mexico: .... Not Applicable

www.natureserve.org

Arctic Loon

Scientific: *Gavia arctica*
Français: Plongeon arctique
México: Colimbo Ártico
Order: Gaviiformes
Family: Gaviidae

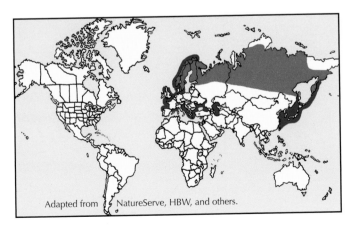

An interesting case of a species pair in which one species breeds mostly in the Old World while the other breeds mostly in the New World is that of the Arctic Loon and the Pacific Loon. The name "Arctic Loon" used to include both forms, but the species was split because the breeding ranges overlap with apparently little or no hybridization where they co-occur in northeastern Asia and western Alaska (Banks et al. 1985). Perhaps the use of different habitats within this region is the condition allowing the two to coexist. Apparently, Arctic Loons breed more inland in the forest belt and tundra, while Pacific Loons breed closer to the coast (Russell 2002). If this is indeed the case, the two species may have little opportunity to meet and reproduce during the breeding season. The occurrence of birds with seemingly inter-species intermediate characteristics suggests that hybridization may occur, though perhaps infrequently. The Arctic Loon has been studied fairly well in Europe and western Eurasia, but very little in western North America, so uncertainty exists as to which findings from the former region apply in the latter, where the subspecies (the "Green-throated Loon", *G. a. viridigularis*) is different (Russell 2002).

Arctic and Pacific Loons can be challenging to differentiate in the field. In all plumages (e.g., breeding, nonbreeding, juvenal), Arctic Loons have a white upward flaring rump patch, but Sibley (2000) warns that: the amount of white varies with posture; other loon species may

Bosse Haglund

roll sideways, exposing a white belly; and, injured or oiled birds often lean to one side. Arctic Loons are larger and have a less rounded head, longer neck, and thicker bill held at a higher angle than do Pacific Loons. In breeding plumage, the nape is darker in Arctic Loons and the black and white neck stripes are more pronounced, lining up with the breast lines. Arctic Loons have lower-pitched voices than Pacific Loons. The color of the throat patch is difficult to see.

Adapted from NatureServe, HBW, and others.

NatureServe
Conservation Status

Global: .... Secure
Canada: .... Not Applicable
US: ....Critically Imperiled
Mexico: .... Not Applicable

www.natureserve.org

Pacific Loon

Scientific: *Gavia pacifica*
Français: Plongeon du Pacifique
México: Colimbo Pacífico
Order: Gaviiformes
Family: Gaviidae

"The Pacific Loon is noted for its peculiarly loud, weird, and prolonged, shrill scream" (McFarlane 1891, 416). It has "a habit of getting off alone in some small pond and howling like a fiend for upward of half an hour at a time. It is a most bloodcurdling, weird, and uncanny sort of a scream, and the amount of noise they make is something wonderful. They can be heard for miles" (Murdoch 1885). The other four loon species also make odd, loud calls, often at night. Hence the common expression, "crazy as a loon".

As in other birds, loons have much to say to each other. They communicate: their territoriality to neighbors and potential intruders; aggression to challengers; desire to shift incubating duties to their mate; warnings of danger to their chicks; and recognition to their neighbors. The four larger species, including the Pacific Loon, have four basic calls (yodel, wail, tremolo, hoot) with variations, but they also croak, purr, and yelp. Low-pitched Common Loon yodels warn other males that the bird is heavy and in good enough condition to win a territorial battle—which may become a fight to the death (Evers et al. 2010). Perhaps this also occurs in other loon species including Pacific Loons. Loon vocalizations are often used interchangeably, so it is easier to understand the meaning of a particular vocalization emitted at a given time and place if you have spent some time on the lake and know something of the context.

A pair nested on this artificial island in Anchorage, Alaska. The roof helps keep away Mew Gulls.

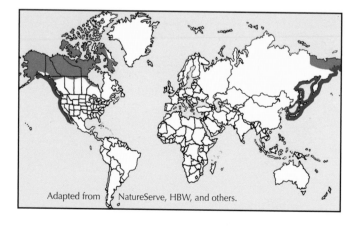

Adapted from NatureServe, HBW, and others.

NatureServe
Conservation Status

Global: .... Secure
Canada: .... Secure
US: .... Secure
Mexico: .... Not Applicable

www.natureserve.org

Common Loon

Scientific: *Gavia immer*
Français: Plongeon huard
México: Colimbo Común
Order: Gaviiformes
Family: Gaviidae

Do you remember all the concern about acid rain in the 1980s? Long Point Bird Observatory hired me to find out how Common Loons, at the top of the food chain, might be affected by lake acidification. From 1982 to 2007, I surveyed dozens of single-pair lakes (pH: 4.0 to 8.5) near Sudbury, Ontario, the hardest hit area in North America. No chicks fledged on lakes with pH < 4.4 regardless of lake size, but they did on lakes with higher pH if the lakes were large. Why? Food for chicks declines on acidic lakes, but the larger the lake, the more food there is for one pair to feed their young. Over the 25 years, as acidic lakes became less so due to reduced sulphur dioxide emissions from the Sudbury smelters and lower sulphur deposition from long-range sources, some lakes that had been too acidic later produced chicks. These results, combined with observations made of parents feeding their young, showed that loon reproduction drops when pH is 6.0 or lower (Alvo 2009). Hundreds of thousands of lakes with pH < 6.0 still exist in northeastern North America because of pollution from vehicles and coal-fired power plants. The "triple whammy" of acidification, global warming, and increased ultraviolet radiation (Schindler 1998) warms and alters these boreal lakes, while invasive animal and plant species advance and mercury levels become magnified on their way up the food chain. Unfortunately, public interest in, and scientific research on, lake acidification, have declined. I hope to entice other scientists to extend and deepen my study to follow the recovery from acidification in Sudbury while the other changes unfold.

Chases may end in the murder of a territorial male by a challenging male.

Parents feed their young by diving, capturing a food item, surfacing, then transfering the item to the chick.

Egg with two small holes in the center made by a predator, likely a bird.

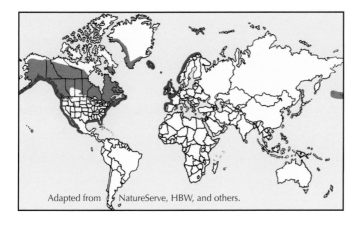

Adapted from NatureServe, HBW, and others.

NatureServe
Conservation Status

Global: .... Secure

Canada: .... Secure

US: .... Apparently Secure

Mexico: .... Not Applicable

www.natureserve.org

Yellow-billed Loon

Scientific: *Gavia adamsii*
Français: Plongeon à bec blanc
México: Colimbo Pico Amarillo
Order: Gaviiformes
Family: Gaviidae

Yellow-billed Loons and Common Loons are sister species and form a superspecies. This superspecies includes no other species, so the two are also a species pair. In the breeding season, they co-occur only in a narrow part of northwestern North America, but are so closely related that if they were brought together by the breaking down of geographical barriers, they would likely interbreed and produce fertile offspring. Sister species start as one species, then become geographically isolated despite some possible range overlap. The superspecies isn't an established category in the taxonomic classification system, but rather a convenient way to show that species are very closely related, though not closely enough to be considered subspecies (Pettingill 1985).

The raft-shaped bodies of loons make these birds stable on the water surface such that there is little tendency to tip over sideways. Their specific gravity is near that of water. They can increase it by expelling air from their feathers and from internal air sacs, and thus sink slowly and quietly. Birds that dive from the water surface are usually heavier than equal-sized landbirds because their bones contain less air, while landbirds are built lighter to enhance their flying skills. In order to extend their time under water during dives, loons tap into residual air in the air sacs and oxygen in the muscles. The diving bird champions, however, using other physiological tricks (not all completely understood), are penguins (McGowan 2004).

Edgar T. Jones

Christian Artuso

Adapted from NatureServe, HBW, and others.

NatureServe
Conservation Status

Global: ...... Apparently Secure
Canada: ..... Apparently Secure
US: .... Vulnerable
Mexico: .... Not Applicable

www.natureserve.org

Least Grebe

Scientific: *Tachybaptus dominicus*
Français: Grèbe minime
México: Zambullidor Menor
Order: Podicipediformes
Family: Podicipedidae

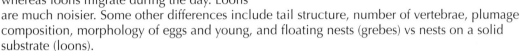

Both loons and grebes dive from the water surface, but they are not closely related. In fact, each group comprises a separate taxonomic order. Grebes have lobed toes with partial webbing rather than the complete webbing of loons. Grebes migrate at night, whereas loons migrate during the day. Loons are much noisier. Some other differences include tail structure, number of vertebrae, plumage composition, morphology of eggs and young, and floating nests (grebes) vs nests on a solid substrate (loons).

Of the seven grebe species that breed in North America, the Least Grebe is the only strictly tropical-breeding one, residing south to central South America. Our smallest grebe, it often uses little bodies of water, including roadside ditches and farm reservoirs. Where it occurs with the Pied-billed Grebe, it occupies shallower ponds (Ridgely and Gwynne 1989). The use of small ponds makes Least Grebes vulnerable to humans, but they move to other bodies of water if disturbed (Storer 2011). Unlike other grebes, Least Grebes are hunted in much of their range.

Grebes are often called "water witches". The Least Grebe's generic name *Tachybaptus* means "quick diver". Colorful descriptions exist of the amazingly quick diving abilities of various aquatic diving birds. "On the slightest alarm [Least Grebes] dive with the quickness of thought, and so vigilant is their eye and so rapid their motion that ordinarily the fowling piece [a shotgun] is discharged at them in vain" (Gosse 1847, 441). When he was concealed, however, Gosse

Larry Master

killed them easily, concluding, "their quick eye detects and takes alarm at the small but sudden motion of the falling hammer". Bent (1919, 56) didn't "dare to say for how long a time the loon succeeded in dodging their well-directed shots, or how many cartridges were wasted". "One may shoot twenty times and misse, for seeing the fire in the panne, [Great Cormorants] dive under the water before the shot comes to the place where they were" (Wood 1634, 33).

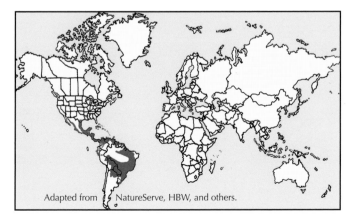

Adapted from NatureServe, HBW, and others.

NatureServe
Conservation Status

Global: .... Secure
Canada: .... Not Applicable
US: .... Vulnerable
Mexico: .... Secure

www.natureserve.org

Pied-billed Grebe

Scientific: ***Podilymbus podiceps***
Français: Grèbe à bec bigarré
México: Zambullidor Pico Grueso
Order: Podicipediformes
Family: Podicipedidae

Grebe chicks look like little prisoners with striped coveralls. This patterning gradually disappears while they mature over their first calendar year (Cramp 1977). No simple explanation exists for striping in grebe chicks vs dull coloring in loon chicks (Kilner 2006).

These photos of the Pied-billed Grebe, the only one of our seven grebes with a ringed bill, offer a glimpse at the secret lives of this often-ignored group of loon-like diving birds.

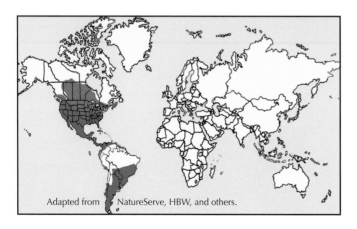

Adapted from NatureServe, HBW, and others.

NatureServe
Conservation Status

Global: .... Secure
Canada: .... Secure
US: .... Secure
Mexico: .... Secure

www.natureserve.org

Horned Grebe

Scientific: *Podiceps auritus*
Français: Grèbe esclavon
México: Zambullidor Cornudo
Order: Podicipediformes
Family: Podicipedidae

The Horned Grebe (or "Slavonian Grebe" in much of the Old World) eats mainly aquatic arthropods in summer, and mainly fish, amphipods, and polychaete worms in winter. Grebes are the only birds that regularly eat feathers. Parents feed them to chicks. Early naturalists thought that feathers muffled the movements of live prey in the stomach, or that feathers kept the stomach comfortably full after food had passed into the intestine.

Organisms eaten by grebes often have hard parts and may contain parasites. Swallowed food passes down the esophagus to the proventriculus, where it receives gastric juices. The second part of the stomach, the muscular gizzard, breaks down the food mechanically. Feathers may protect stomach linings from spiny fish bones. The stomach contents, including feathers, are often found in two bundles: the main bundle in the gizzard, and a small plug in the gizzard's pyloric exit to the small intestine. Feathers in the main bundle may contribute substance to the stomach, allowing pellets to be regurgitated regularly (e.g., every two days) along with parasites and undigestible hard parts such as arthropod chitin. These feathers may also separate hard food items, thus increasing their surface area exposed to digestive stomach acid and hastening digestion. The pyloric plug filters the stomach contents, preventing the passage of hard parts to the small intestine and giving digestible ones such as fish bones more time to be broken down.

Each of these mechanisms may apply more to some grebe species than to others. In Horned Grebes, feathers are more abundant in stomachs of birds from areas where fish is a staple food than from areas where fish are not eaten. Regurgitated pellets in this species are roughly 2 cm long and 0.65 cm in diameter (Stedman 2000). Grebes shed feathers throughout the year, thus helping provide a regular feather supply for digestion.

Edgar T. Jones

Dennis Paulson

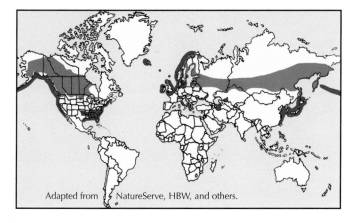
Adapted from NatureServe, HBW, and others.

NatureServe
Conservation Status

Global: .... Secure

Canada: .... Secure

US: .... Secure

Mexico: .... Not Applicable

www.natureserve.org

Red-necked Grebe

Scientific: *Podiceps grisegena*
Français: Grèbe jougris
México: Zambullidor Cuello Rojo
Order: Podicipediformes
Family: Podicipedidae

"As soon as they are able to feed and to swim about, the young may be seen riding in safety on their mother's back as she swims about the lake, clinging to her plumage when she dives and coming to the surface with her as if nothing had happened" (Bent 1919, 12). Given that grebes take almost all their food under water, it is likely that submerging is one of the first things that chicks must learn to do (or tolerate) before learning how to feed. An underwater ride on a parent's back may be that chick's first lesson in the ways of grebe feeding.

In all animals, methods of food capture, methods of food handling, and diet differ in adults vs juveniles. In fish, amphibians, and insects, this variation may result from some great difference between the adults and young such that nature treats them as distinct organisms. (Think of a caterpillar vs a butterfly.) In birds, though, pressure seems to act on juveniles to develop into adults as fast as possible. The main constraints keeping young birds from rapidly becoming adults are physiological (an egg can't instantaneously become an adult) and the time required to learn how to feed like an adult (Marchetti and Price 1989). So when you see a Red-necked Grebe dive with a chick on its back, perhaps consider that the chick is hanging on in the race to grow into, and learn quickly about, the world of adult feeding. It is an apprenticeship.

Edgar T. Jones

Dennis Paulson

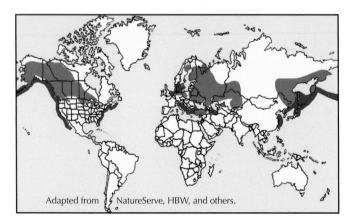
Adapted from NatureServe, HBW, and others.

NatureServe
Conservation Status

Global: .... Secure

Canada: .... Secure

US: .... Secure

Mexico: .... Not Applicable

www.natureserve.org

Eared Grebe

Scientific: ***Podiceps nigricollis***
Français: Grèbe à cou noir
México: Zambullidor Orejón
Order: Podicipediformes
Family: Podicipedidae

Freshwater marsh nests of Eared Grebes can be only half a meter apart. Thousands of these tame, colonial birds were easily shot in Oregon and California. "The breasts were stripped off, dried, and shipped to New York, where they were much in demand for ladies' hats, capes, and muffs" (Bent 1919, 33).

This is the last North American migrant to move to its wintering areas. After breeding, most New World birds move to Mono Lake, California, or to Great Salt Lake, Utah, to gorge themselves on brine shrimp (*Artemia*) and brine flies (*Ephedra*) in these hypersaline (2.5 times the salinity of seawater, and pH 10) habitats that lack fish predators.

While molting there, they transform into eating machines: their mass more than doubles, the flight muscles atrophy, and the digestion and food-storage organs expand. These physiological changes are the most extreme known in birds. They are then reversed to allow the grebes to become flying machines again for a nonstop flight to more southerly wintering areas. When finished eating there, they use up their fat to increase heart size, reduce digestive organ mass, and migrate soon after food supplies run out. Repeating this physiological cycle 3–6 times each year, the Eared Grebe has the longest flightless period (perhaps 9–10 months of the year) of the world's birds that can fly.

This species never drinks fresh water at Mono Lake. Rather, it maintains its salt balance by using its fleshy tongue like a baleen whale does to crush prey against the palate and remove the salt water. Also, its food is less saline than Mono Lake's water (Mahoney and Jehl 1985; Cullen et al. 1999).

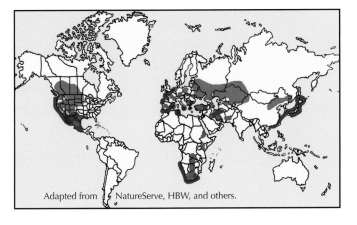

Edgar T. Jones

Adapted from NatureServe, HBW, and others.

NatureServe
Conservation Status

Global: .... Secure
Canada: .... Secure
US: .... Secure
Mexico: ..... Apparently Secure

www.natureserve.org

Western Grebe

Scientific: *Aechmophorus occidentalis*
Français: Grèbe élégant
México: Achichilique Pico Amarillo
Order: Podicipediformes
Family: Podicipedidae

Courtship ceremonies are well-known in birds (and other animals), but those performed by the Western Grebe and Clark's Grebes are very complex. The Western Grebe's is "the most significant wedding dance I have ever seen in bird life.… As two birds were swimming together, both dove. They rose to the top of the water a few moments later, each holding a piece of moss or weed in the bill. Instantly they faced each other and rose, treading water, with bodies half above the surface and necks stretched straight up. They treaded around, breast to breast, until they made three or four circles, and then dropped down to a normal attitude, at the same time flirting the moss out of their mouths and swimming off in an unconcerned manner.… I saw it three times within close range, and each time it was exactly the same" (Finley, W. L., in Bent 1919, 2). It is as if the birds had memorized and performed it repeatedly. This "play" has been reenacted for millenia, but at some point in evolution some individuals developed an olive-yellow bill and more black around the eyes, while others had an orange-yellow bill and less of that black. Advertising calls involved two notes in the former birds, but only one note in the latter. These subtle differences went largely unnoticed by researchers until they found that individuals mated only with similar-looking and similar-sounding individuals. Thus, two species were involved—the Western Grebe and the Clark's Grebe.

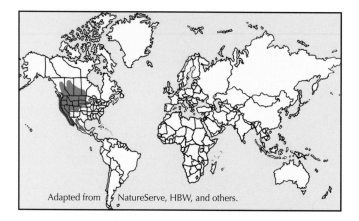

Dennis Paulson

Female grebes mount males often enough during courtship (27% in some species) that biologists studying species with identical-looking sexes (e.g., grebes, loons, boobies), should be careful when relying on mounting behavior to determine sex. Such "reverse mounting" isn't aberrant behavior, but simply a fascinating part of normal courtship (Nuechterlein and Storer 1989).

Adapted from NatureServe, HBW, and others.

NatureServe
Conservation Status

Global: .... Secure
Canada: .... Secure
US: .... Secure
Mexico: .... Secure

www.natureserve.org

Clark's Grebe

Scientific: *Aechmophorus clarkii*
Français: Grèbe à face blanche
México: Achichilique Pico Naranja
Order: Podicipediformes
Family: Podicipedidae

The Clark's Grebe looks like the Western Grebe and was treated as one of its color morphs until significant differences were found, not only in their morphology and advertising call, but also in their foraging behavior and DNA.

Grebes and loons regularly carry small chicks on their backs to brood and protect them from predators. Some waterfowl also do so to some degree (Johnsgard and Kear 1968). Lucky observers have seen waterfowl carrying chicks from high nest sites down to the water, including a Ruddy Shelduck (*Tadorna ferruginea*) with the young tucked between its neck and shoulder. Other species of birds also carry their young at times. For example, an alarmed African Jacana (*Actophilornis africana*) crouched and allowed its three chicks to run under its wings, then clasped the chicks and walked away with them, whereas an adult Spotted Sandpiper flew for a distance with a chick clutched between its legs. The Eurasian Woodcock may carry chicks between its legs and breast (Terres 1991).

Clearly, the grebes and loons are the avian chick-carrying masters of North America, doing this on a regular basis, but the male Sungrebe beats even them with its chick-carrying "pocket" complete with feather "seat-belt" shroud under each wing (Stiles and Skutch 1989).

Carol Blackard

The Clark's Grebe (background) has an orange bill, and eye on white; the Western Grebe (foreground) has a yellow bill, and eye on black.

The Red-necked Grebe abandons its nest at night for 3–9 hours without ill effect on the eggs, even though egg temperature drops. Nuechterlein and Buitron (2002) suggest that adults do so proactively to avoid predation to themselves or to their eggs. Do Clark's Grebes, other grebes, or even other birds also abandon their eggs to deal with the age-old quandary of night survival?

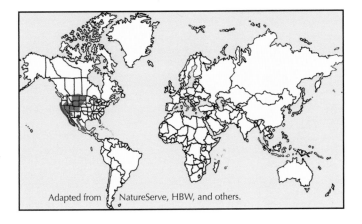

Adapted from NatureServe, HBW, and others.

NatureServe
Conservation Status

Global: .... Secure

Canada: .... Imperiled

US: .... Secure

Mexico: .... Secure

www.natureserve.org

American Flamingo

Scientific: *Phoenicopterus ruber*
Français: Flamant des Caraïbes
México: Flamenco Americano
Order: Phoenicopteriformes
Family: Phoenicopteridae

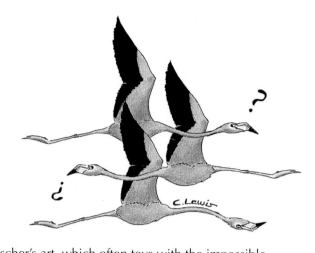

"Flamingos in flight resemble no other bird known to me. With legs and neck fully outstretched, and the comparatively small wings set halfway between bill and toes, they look as if they might fly backward or forward with equal ease" (Chapman 1908b, 181). This description conjures up images of M. C. Escher's art, which often toys with the impossible.

The American Flamingo is Critically Imperiled in the US (N1N), the second "N" referring to its nonbreeding status as a rare visitor to the southern US. It could equally be ranked as Possibly Extipated (NH or NHB) as a breeder in the US (the "B" in "NHB" referring to "breeding") because it probably bred with some regularity in southern Florida; thus it is treated in the main part of this book. Bent (1926, 2) wrote that, "The flamingo is no longer to be found, except possibly as a rare straggler, on the North American continent, but in Audubon's time [the early-mid 1800s] it was fairly abundant in extreme southern Florida.... It was supposed to breed somewhere in that vicinity, but the breeding grounds were never found".

The four main breeding colonies are in Mexico, Cuba, the Bahamas, and the Netherlands Antilles. The taxon is sometimes treated as the same species as the Greater Flamingo of the Old

Courtesy of the Edgar T. Jones Collection, Provincial Museum of Alberta

World (*Phoenicopterus roseus*), but the American Ornithologists' Union now treats it separately based on differences in color of plumage and bill, and in displays and vocalizations (Banks et al. 2008).

Flamingos are large waders with long necks and legs. Their downcurved bill is used for filter-feeding. Identification can be challenging because escapees of all six of the world's species have been observed in North America. Also, pale washed-out birds may be immatures or escapees. The pink color often fades in captivity, particularly if the birds aren't given a proper diet (Dunn and Alderfer 2006; Peterson 2008).

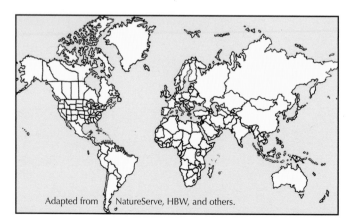

Adapted from NatureServe, HBW, and others.

NatureServe
Conservation Status

Global: ...... Apparently Secure
Canada: .... Not Applicable
US: ....Critically Imperiled
Mexico: ....Critically Imperiled

www.natureserve.org

Northern Fulmar

Scientific: *Fulmarus glacialis*
Français: Fulmar boréal
México: Fulmar Norteño
Order: Procellariiformes
Family: Procellariidae

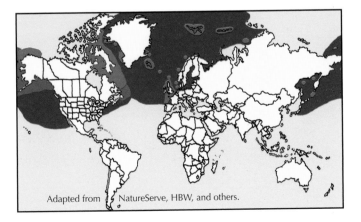

Procellariiformes means "pertaining to storms". This taxonomic order includes the albatrosses, diving-petrels, fulmars, petrels, prions, shearwaters, and storm-petrels. Also known as "tubenoses", these pelagic birds have one or two tubes on the bill that enclose the nostrils. Excess salt, obtained by drinking seawater, is excreted through these tubes.

The Northern Fulmar looks like a gull that uses stiff-winged flapping as it glides close to the ocean surface. It has a thick neck, large head, and stout bill. A bird of the Arctic (the species name *glacialis* means "icy"), it is particularly active in rough and stormy seas. "The Northern Fulmar is the constant companion of the Arctic whalers and is well known to the hardy explorers who risk their lives in dangerous northern seas, where it follows the ships to gorge itself on what scraps it can pick up" (Bent 1922, 31).

It has expanded greatly in range and numbers over the past 250 years, probably as a result of the increased availability of food, especially fish offal, from commercial fishing. Other factors likely also have contributed. It is still increasing in some areas, though more slowly (Carboneras et al. 2014). The Northern Fulmar probably does much of its foraging at night, relying on its sense of smell to find food. It also apparently uses this sense to avoid petroleum floating on the water surface; it deliberately avoided settling on waters that had been experimentally polluted with petroleum, and, relative to other species, only a few birds were killed in the Exxon Valdez oil spill of 1989 in Alaska.

Dennis Paulson

John Hoyt

Fulmars are among the longest-lived birds, some surviving more than 50 years in the wild. Reproduction is slow—most birds don't start breeding until they are 8–10 years old, and females produce only one egg per year (Mallory et al. 2012).

Adapted from NatureServe, HBW, and others.

NatureServe
Conservation Status

Global: .... Secure

Canada: .... Secure

US: .... Secure

Mexico: .... Not Applicable

www.natureserve.org

Manx Shearwater

Scientific: ***Puffinus puffinus***
Français: Puffin des Anglais
México: Pardela Boreal
Order: Procellariiformes
Family: Procellariidae

An Old World species, the Manx Shearwater's occurrence in the northwest Atlantic Ocean was accidental before 1900. It reputedly bred in Bermuda before 1905, but was later extirpated. Around 1950, more birds were being seen offshore of eastern North America.

The first documented North American nesting occurred in 1973 at Penikese Island, Massachusetts, but the only known site with regular breeding in North America is Middle Lawn Island, Newfoundland, where breeding was first documented in 1977. The species has bred elsewhere on the east coast of North America, but apparently not regularly. Roul (2010), using acoustic recording devices for this nocturnal species, found evidence of possible breeding on three other Newfoundland islands.

In this species both parents feed the sole chick for up to 60 days, traveling separately up to 330–380 km away from the colony to forage, then returning with partially digested food to regurgitate to the chick. They desert the chick before it can fly so that they can make their own long trans-equatorial migration to the south Atlantic Ocean and molt. The chick then fasts for 11–15 days in its burrow, losing up to a third of its weight before fledging. It comes out only at night to exercise its wings, and may walk up to 45 m from the burrow, sometimes associating with its peers.

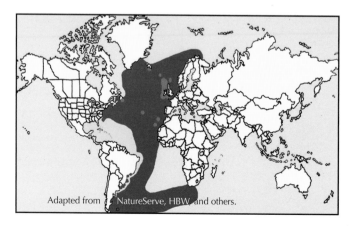

Gail Fraser

It is thought that once ready, the first-year bird makes its way south to its parents' wintering area. The more that the bird weighs, the greater the chance it will survive the long migration. In fact, with enough stored fat, it may make the entire journey without stopping. Two- to five-year-old birds visit their natal colony and start prospecting for burrows in which to nest. Some of them nest very close to the burrows in which they hatched long before (Lee and Haney 1996).

Adapted from NatureServe, HBW, and others.

NatureServe
Conservation Status

Global: .... Secure

Canada: .... Vulnerable

US: .... Critically Imperiled

Mexico: .... Not Applicable

www.natureserve.org

Fork-tailed Storm-Petrel

Scientific: *Oceanodroma furcata*
Français: Océanite à queue fourchue
México: Paíño Gris
Order: Procellariiformes
Family: Hydrobatidae

The Fork-tailed Storm-Petrel uses its sense of smell to find food, taking crustaceans and fish from the ocean's surface while hovering or settling briefly. It also follows boats and feeds on their waste. At sea, it is often seen feeding individually or in small flocks of fewer than 10 individuals, but a dead marine mammal may attract a few hundred birds. As do other tubenoses, it transforms its food into oil and stores it in its proventriculus. This oil is very light and is thought to be an efficient means of transporting energy-rich food back to the nest, which can be hundreds of kilometers from the foraging grounds. Oil production and storage aren't limited to breeding birds. Because this species' food is often found in surface slicks, it probably ingests pollutants such as petroleum. However, it may be less susceptible than other seabirds to petroleum toxicity because its diet contains large amounts of "n-hexanes", which are chemically similar to petroleum. The oil it produces can also be used as a weapon in that it is expelled from the mouth and nostrils at a predator. If the oil is orange or yellow, the bird has likely eaten crustaceans, whereas milky or light oil results from the digestion of fish. Chicks are well-adapted to their parents' unpredictable visits to the nest that result from foraging areas being far away and patchily distributed. For example, chicks raise their body temperature and grow when fed, but while they are waiting for hours or days to be fed, they decrease their body temperature and metabolic rate and may become torpid (Dee Boersma and Silva 2001).

Gavin Bieber

Kelly A. Boadway

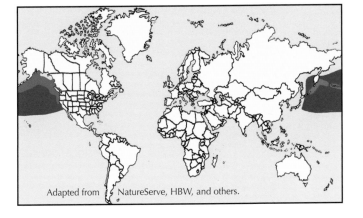

Adapted from NatureServe, HBW, and others.

NatureServe
Conservation Status

Global: .... Secure
Canada: ..... Apparently Secure
US: .... Secure
Mexico: .... Not Applicable

www.natureserve.org

Leach's Storm-Petrel

Scientific: *Oceanodroma leucorhoa*
Français: Océanite cul-blanc
México: Paíño de Leach
Order: Procellariiformes
Family: Hydrobatidae

Imagine camping in the middle of a petrel colony without knowing it. The ground below you would be riddled with burrows containing nests with incubating birds or lone young, and you could easily overlook the entrances, which are usually surrounded by vegetation. "It is a weird experience to spend a night in a petrel colony during the breeding season. Night is their season of activity, birds coming and going all the time..., hardly discernible in the darkness, uttering their loud and peculiar cries as they call to or greet their mates.... It is a wonder that the incoming birds can find their mates or their burrows in the darkness and the confusion of thousands of fluttering birds" (Bent 1922, 141). Experiments show that they use their sense of smell. "There was nothing ... to indicate that beneath our feet lay a buried city ... teeming with life, a city of storm waifs gathered from an expanse of 1000 watery leagues" (Dawson 1909, 875). Most people never see petrels, but birders do so on specially organized pelagic boat trips.

Biologists working on the Leach's Storm-Petrel probably obtain the best appreciation of this species' "many secrets [that] remain hidden in underground burrows and vast ocean spaces" (Huntington et al. 1996). Researcher Kelly A. Boadway was most impressed with "Everything. Where they nest (offshore islands), where they live (open ocean), the fact that they can live there and be such small birds, how long the chicks can go between feedings, nocturnal activity at the colony, biparental care, their endearing calls, and the great density of birds at the colony".

Kelly A. Boadway

While chicks, incubating adults, and brooding adults sleep in burrows, other parents and nonbreeding adults are thought to sleep out at sea. Not all burrows in a colony are used for nesting. Adults may visit these burrows, but it is unknown what exactly they are doing when hidden underground (Huntington et al. 1996).

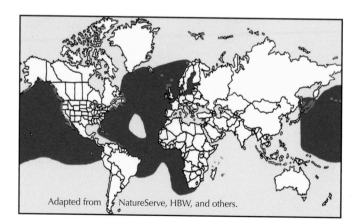

Adapted from NatureServe, HBW, and others.

NatureServe
Conservation Status

Global: .... Secure

Canada: .... Secure

US: .... Secure

Mexico: ..... Apparently Secure

www.natureserve.org

Ashy Storm-Petrel

Scientific: *Oceanodroma homochroa*
Français: Océanite cendré
México: Paíño Cenizo
Order: Procellariiformes
Family: Hydrobatidae

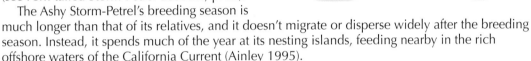

"The strong musky odor of the petrels renders their discovery in the rock piles easy. It is only necessary to insert the nose into likely crevices to find them. With little practice one may become very expert in this kind of hunting, readily determining whether it is an auklet or a petrel that has its residence in any particular cranny" (Loomis 1896). This strong odor emanates from the stomach oil that parents regurgitate for the chick (see Fork-tailed Storm-Petrel account, p. 93).

The Ashy Storm-Petrel's breeding season is much longer than that of its relatives, and it doesn't migrate or disperse widely after the breeding season. Instead, it spends much of the year at its nesting islands, feeding nearby in the rich offshore waters of the California Current (Ainley 1995).

This species-at-risk has a very limited global range and is endemic as a breeder to California, except for one island off the coast of Baja California, Mexico. The global population is estimated at only 7200 individuals that breed on six island groups and on three groups of offshore rocks. About half of the population breeds in one colony on the Farallon Islands off the coast of San Francisco, California. A major population decline has occurred over the past few decades. Fortunately, all US nesting areas are protected. Threats to the species include potential oil spills, predation (especially from increasing Western Gull populations at breeding colonies, but also from introduced rats and mice), lighting from boats and offshore energy platforms, plastic ingestion, eggshell thinning from the uptake of organochlorines, and climate change. The Ashy Storm-Petrel is thus Imperiled globally and in the US and Mexico (NatureServe 2014b).

Annie Schmidt

Sara Acosta, Point Reyes Bird Observatory Conservation Science and US Fish and Wildlife Service

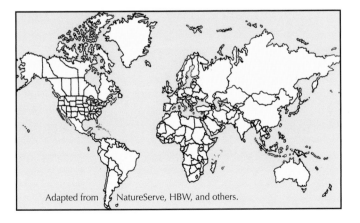

Adapted from NatureServe, HBW, and others.

NatureServe
Conservation Status

Global: ....Imperiled
Canada: .... Not Applicable
US: ....Imperiled
Mexico: ....Imperiled

www.natureserve.org

Black Storm-Petrel

Scientific: *Oceanodroma melania*
Français: Océanite noir
México: Paíño Negro
Order: Procellariiformes
Family: Hydrobatidae

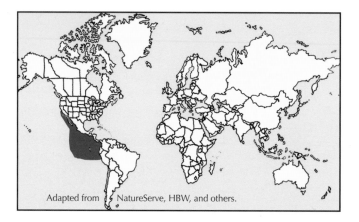

I'M TIRED OF ALWAYS RUNNING LARRY! I FEEL LIKE A CRIMINAL...

The genus name *Oceanodroma* means "ocean running", referring to the practice, when these storm-petrels are foraging, of hovering while the feet push off the water surface. When flying over the ocean for travel, however, the Black Storm-Petrel is slow and deliberate compared to other storm-petrels, using deep wing-beats and long glides like a nighthawk.

When migrating to its wintering grounds, some birds move north while others go south, as do some other seabird species nesting in the Gulf of California. The movement north precedes the southward one, but it isn't known whether the same individuals are involved. The Heermann's Gull and Craveri's Murrelet, in contrast, move only northward from the Gulf. All these movements away from the Gulf of California seem to be directed and timed to allow avoidance of the region's well-known hurricanes that pass through during late summer and early fall.

The Black Storm-Petrel nests on the same islands as do other storm-petrels, shearwaters, pelicans, cormorants, gulls, murrelets, and auklets. It also competes for, or shares nesting cavities with, three alcid species that tend to nest earlier in the season. The Black Storm-Petrel arrives at its colony from the ocean after dark while diurnal predators such as Western Gulls are no longer on the wing. Its aerial activity peaks soon afterward, but then drops off gradually until midnight. A second period of calling and apparent aerial activity of unknown function peaks

Annie Schmidt

soon after 2:00 a.m. followed by a rapid decline at 4:00 a.m. when the birds return to sea. A bright moon and clear sky result in fewer birds returning to the colony at night while gulls are much more active. The only defenses against mammalian predators are the use of precipitous nest sites and the presence of *Cholla* cactus and other spiny plants. This can be a double-edged sword, however, because impalement on plant spines by the Black Storm-Petrel is known to occur. Peregrine Falcons represent another danger— storm-petrels are safer when resting on water than when they are in the air (Ainley and Everett 2001).

Adapted from NatureServe, HBW, and others.

NatureServe
Conservation Status

Global: .... Vulnerable

Canada: .... Not Applicable

US: .... Critically Imperiled

Mexico: .... Vulnerable

www.natureserve.org

Wood Stork

Scientific: ***Mycteria americana***
Français: Tantale d'Amérique
México: Cigüeña Americana
Order: Ciconiiformes
Family: Ciconiidae

The "flinthead" is North America's only breeding stork. It is a tactile feeder, capturing food using its sense of touch. This allows it to forage in wetlands with murky water. Walking slowly through water with its bill partly open and submerged, it moves its bill from side to side, snapping it shut when it feels prey. (When blindfolded, captive Wood Storks forage just as efficiently as birds without visual restriction.) They purposefully startle prey by pumping one foot up and down about five times, then repeating this process with the other foot. Another method of startling prey is to partially open one wing and hold it open for a few seconds with the tips of the primary feathers low over the water surface. Fledglings perform all the foraging habits of adults, but often go to inappropriate places without prey, such as flooded lawns and rain puddles. Wood Storks like to forage in groups in open wetlands, moving through water in a line. They also feed with other avian waders, especially when prey density is high (Coulter et al. 1999).

David Laliberte

Southern Florida used to have extensive wetlands with large Wood Stork breeding colonies. The Kissimmee River to the north was the main water source. When it was channelized after World War II to control flooding for human interests, it became stagnant, and about 14,000 ha (35,000 acres) of floodplain wetlands vanished. The entire ecosystem was profoundly altered to the disadvantage of many of its residents, and the large Wood Stork colonies perished. In 1984, the US population was legally listed as "Endangered". Reversal of Florida's great hydrological error began in the 1990s with the initiation of the very expensive ecosystem-level Kissimmee River Restoration Project and the Everglades Restoration Project (Podulka et al. 2004).

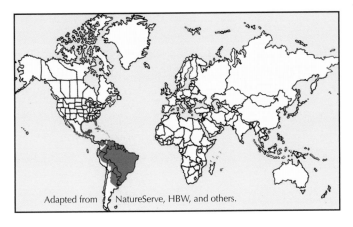

Adapted from NatureServe, HBW, and others.

NatureServe
Conservation Status

Global: ...... Apparently Secure
Canada: .... Not Applicable
US: .... Vulnerable
Mexico: ..... Apparently Secure

www.natureserve.org

Magnificent Frigatebird

Scientific: *Fregata magnificens*
Français: Frégate superbe
México: Fragata Tijereta
Order: Suliformes
Family: Fregatidae

The Magnificent Frigatebird has the greatest wing-area to body-weight ratio of all living birds. It glides for hours, hardly flapping, and withstands hurricanes. Also called "Man-o'-War Bird", it steals food from other birds or finds its own food, but it never swims or dives. It has been seen 1000 km from land, and its short legs and small feet don't allow for walking.

Although it hasn't been proven, the Magnificent Frigatebird sexes appear to breed on different cycles, the male every year and the female every second year. In no other seabird is this thought to happen. We do know that after sharing in incubation and chick-feeding duties equally with the female, the male Magnificent Frigatebird departs when the sole chick is only half grown, and we know that this would allow the male to breed with another female while his initial mate continues to care for the chick through its first year. It is unclear whether males do in fact breed every year (with different females) or whether they just skip a year after breeding.

Before being found nesting in the Florida Keys in 1969, the Magnificent Frigatebird was considered only a visitor to North America (Diamond and Schreiber 2002).

Sequence of photos showing the release of a Magnificent Frigatebird that was oiled during the 2010 Deepwater Horizon disaster, then cleaned and released: just released (left), ready for takeoff (center), liberty (right). See Brown Pelican account (p.109) for photos of the cleaning process that occurs before release. Numerous bird species were oiled and the overall ecosystem effects were catastrophic.

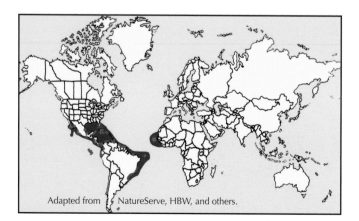

Adapted from NatureServe, HBW, and others.

NatureServe
Conservation Status

Global: .... Secure
Canada: .... Not Applicable
US: .... Not Ranked
Mexico: .... Secure

www.natureserve.org

Brown Booby

Scientific: *Sula leucogaster*
Français: Fou brun
México: Bobo Café
Order: Suliformes
Family: Sulidae

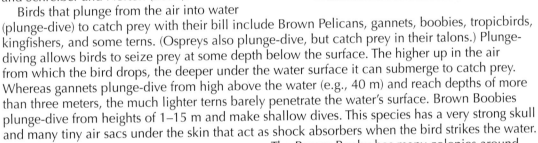

The Brown Booby is a visitor to North America, breeding widely in the West Indies. The famous naturalist J. J. Audubon wrote in 1840 of this species breeding in the Dry Tortugas (Bent 1922), which are seven islets located about 110 km west of Key West, Florida. While some writers have doubted the credibility of Audubon's identification, Bent gave excellent reasons for accepting it. Despite some lingering doubt about its former North American breeding status, I allot it a species account, as have NatureServe (NatureServe 2014b) and Schreiber and Norton (2002).

Birds that plunge from the air into water (plunge-dive) to catch prey with their bill include Brown Pelicans, gannets, boobies, tropicbirds, kingfishers, and some terns. (Ospreys also plunge-dive, but catch prey in their talons.) Plunge-diving allows birds to seize prey at some depth below the surface. The higher up in the air from which the bird drops, the deeper under the water surface it can submerge to catch prey. Whereas gannets plunge-dive from high above the water (e.g., 40 m) and reach depths of more than three meters, the much lighter terns barely penetrate the water's surface. Brown Boobies plunge-dive from heights of 1–15 m and make shallow dives. This species has a very strong skull and many tiny air sacs under the skin that act as shock absorbers when the bird strikes the water.

Peter Sproule

The Brown Booby has many colonies around the world that were last surveyed 30–50 years ago, and new surveys are needed to better assess the species' conservation status. It tolerates researchers well. Some newly-formed pairs may desert eggs if disturbed a few times a day for 1–2 weeks, but researchers stopping at each of the 500 nests at the Johnston Atoll in the North Pacific Ocean to weigh birds and measure eggs caused no desertion (Schreiber and Norton 2002).

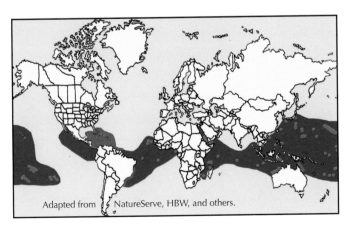

Adapted from NatureServe, HBW, and others.

NatureServe
Conservation Status

Global: .... Secure

Canada: .... Not Applicable

US: .... Possibly Extirpated

Mexico: .... Secure

www.natureserve.org

Northern Gannet

Scientific: *Morus bassanus*
Français: Fou de Bassan
México: Bobo Norteño
Order: Suliformes
Family: Sulidae

The word *fou* in the Northern Gannet's French name means "crazy" and refers to its habit of plunge-diving from a height of 10–40 m at speeds faster than 100 km/h. This species usually fishes in large "frantic-flocks" of up to 1000 birds that concentrate on a school of fish. Plunge-dives take gannets 3–5 m below the water surface.

Both sexes incubate the single egg, but neither parent develops a featherless brood patch on the breast to put the egg in close contact with the warm skin. Rather, the egg is incubated under the webs of the feet, which are replete with blood vessels. Just before the egg is laid, and during incubation, the parents tuck feathers and other nest material around their breast and flanks to reduce heat loss from the nest. The young chick is brooded on the parents' feet for two or three weeks. At the age of about 90 days, the chick moves from its nest to the edge of a cliff. "When the day comes for the mighty plunge to be made, spreading wide its great sails of wings, the young gannet may be seen to half fly, half fall, into the abyss below" (Gurney 1913, 370).

Edgar T. Jones

Nests may be placed on cliffs or on flat ground, often above cliffs.

Once in the air, it avoids the rocks below by traveling 400–800 m and landing on the ocean with a big splash. Apparently it won't be able to take off from the water until three days to two weeks later, yet it begins migrating south by swimming. It may remain away from its natal colony until it is 2–3 years old and spend several years as a nonbreeder, starting to breed when 5–6 years old. Sub-adults and nonbreeding adults in breeding plumage gather in "clubs" to rehearse sexual and territorial behavior, practice landing and taking off from ledges during windy conditions, and learn the routes to nearby foraging areas (Mowbray 2002b).

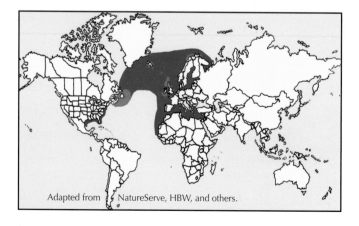

Adapted from NatureServe, HBW, and others.

NatureServe
Conservation Status

Global: .... Secure

Canada: ..... Apparently Secure

US: .... Not Applicable

Mexico: .... Not Applicable

www.natureserve.org

ROBERT ALVO

Brandt's Cormorant

Scientific: *Phalacrocorax penicillatus*
Français: Cormoran de Brandt
México: Cormorán de Brandt
Order: Suliformes
Family: Phalacrocoracidae

Brandt's Cormorants are gregarious in all seasons. They often join feeding flocks with other cormorants, pelicans, gulls, terns, and murres. They also nest in colonies with tubenoses, other cormorants, gulls, and alcids. Aggressive behavior in Brandt's Cormorants is often related to conflicts over nest sites and mates. For example, males arriving at a breeding colony early in the season take the central sites around which later-arriving males nest, but fights may occur. Females fight among each other for a male and his nest site. Regardless of gender, rivals grab each other, push and pull, flap their wings, and tumble over rocks. They may destroy each others' nests or steal nest materials. All this aggressive behavior isn't surprising given that nests are very close together, closer than in any other North American cormorant (2.5 nests/m², Wallace and Wallace 1998), but it must require considerable energy. Other disadvantages of colonial nesting include the increased probability of the spread of diseases and parasites, and interference such as sneak copulations. There must be advantages of living in close proximity that outweigh the disadvantages in this species. For example, nesting colonially may reduce pressure from nest predators in that more individuals are present to watch for them, or perhaps some individuals follow other more knowledgeable ones to good foraging sites. Over evolutionary time, each bird species has arrived at its own solution as to the best nesting density, and this presumably has been done by some weighing, through natural selection, of all the advantages and disadvantages of high vs low densities. This is how such a wide range of nest densities among different species has come to be, from the extreme shown by Brandt's Cormorants to the opposite extreme seen in Common Loons, in which one pair may defend an entire lake.

Parent and chick

Sara Acosta, Point Reyes Bird Observatory Conservation Science and United States Parks Service

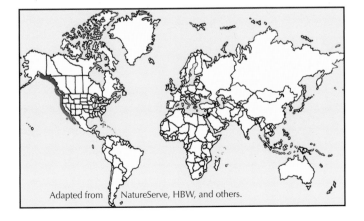

Adapted from NatureServe, HBW, and others.

NatureServe
Conservation Status

Global: .... Secure

Canada: .... Critically Imperiled

US: .... Secure

Mexico: .... Secure

www.natureserve.org

Neotropic Cormorant

Scientific: *Phalacrocorax brasilianus*
Français: Cormoran vigua
México: Cormorán Neotropical
Order: Suliformes
Family: Phalacrocoracidae

As suggested by its current English name, this is the only tropical cormorant of North America. It also stands out from the other species in that it occupies many different wetland habitats in fresh, brackish, and salt water. In coastal marine areas it uses inshore areas including sheltered bays, inlets, estuaries, lagoons, rock outcrops, and islands. Inland, however, it uses broad slow-flowing rivers, fast mountain streams, high elevation Andean lakes, lowland lakes, marshes, and swamps. It nests mostly in live or dead trees, but also uses cliffs, rock outcrops, bare ground, and human structures such as duck blinds. In short, its key habitat requirements are fish, water deep enough for diving, and perches for nesting, roosting, and drying its plumage after feeding.

Juan Bahamon

As in other cormorants the typical foraging method is pursuit-diving from the water surface, but the Neotropic Cormorant is the only cormorant known also to feed by plunge-diving, which likely allows it to include more fish species in its diet. It may forage in strong surf. This species' versatility has allowed it to take advantage of new habitats created by humans, especially geographic-scale reservoirs that store water for human use and/or are used to produce electricity. Since the mid-1970s, it has established inland colonies associated mostly with such reservoirs.

Neotropic Cormorants have been shot by anglers for decades because of the perception that they eat too many sportfish. Cormorants can reduce fish populations in fish farms and small impoundments. It is unknown to what extent such shooting of Neotropic Cormorants affects their North American population, but the US status of Apparently Secure suggests that there is no cause for immediate concern (Telfair and Morrison 2005).

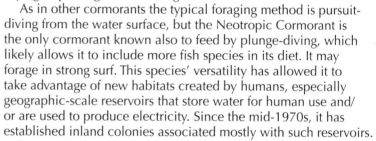

Adapted from NatureServe, HBW, and others.

NatureServe
Conservation Status

Global: .... Secure

Canada: .... Not Applicable

US: .... Apparently Secure

Mexico: .... Secure

www.natureserve.org

Double-crested Cormorant

Scientific: *Phalacrocorax auritus*
Français: Cormoran à aigrettes
México: Cormorán Orejón
Order: Suliformes
Family: Phalacrocoracidae

Big deal.

During the 19th and early 20th centuries, this fish-eating migrant was persecuted throughout its North American range until it disappeared from many colonies. Subsequent protection has allowed it to reoccupy many of its former haunts. People would shoot cormorants because of perceived competition for game fish. Despite considerable research, it still isn't known to what degree this species affects wild fish populations. The other main conflict with humans results from its habit of often nesting in trees.

Excrement emanating from tree nests causes an ionic imbalance in the soil below, eventually killing the trees and other vegetation. Herons often use the same trees for nesting, and when the trees die the heron nests that were built far out on the branches fall to the ground. The cormorants can keep nesting there in subsequent years because their nests are usually closer to the tree trunk and rarely fall down. In the St. Lawrence Estuary and Gulf, the low vegetation that dies below cormorant nests is no longer available as nesting cover for Common Eiders, which breed in large numbers there. Eider down has been an important source of funds used to keep these nesting islands protected from human development (see Common Eider account, p. 43).

Kelly A. Boadway

A Quebec government-sponsored program to help the eiders and to continue protecting the islands from 1989 to 1993 involved shooting adult cormorants, destroying nests, and spraying eggs with an oily substance that asphyxiated the embryos. The female, not realizing that the eggs were dead, would continue to incubate them rather than lay fresh ones. This was only one of numerous programs aimed at controlling this species in North America. A currently-recognized control method is to harass the cormorants at nocturnal roosts using pyrotechnics. This procedure can be used successfully at breeding colonies and at roosts near catfish farms, which are favorite cormorant feeding areas (Hatch and Weseloh 1999).

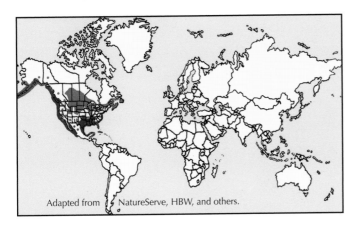

Adapted from NatureServe, HBW, and others.

NatureServe
Conservation Status

Global: .... Secure

Canada: .... Secure

US: .... Secure

Mexico: .... Secure

www.natureserve.org

Great Cormorant

Scientific: ***Phalacrocorax carbo***
Français: Grand Cormoran
México: Cormorán Grande
Order: Suliformes
Family: Phalacrocoracidae

Most of its prey are less than 20 cm long, but the Great Cormorant can capture 1.5 kg fish. For 2–3 millennia, fishers in China have been using trained birds to catch fish to sell. A neck collar prevents the bird from swallowing the fish. A fisher may work with a dozen birds at a time, and several boats of fishers may collaborate. The birds themselves will work together to subdue a very large fish. This cormorant may be able to count to some degree—it is often allowed to swallow the eighth fish it catches (Hatch et al. 2000). Excursions are available in China and Japan for tourists to witness this technique. It was used in England too, though for a different purpose. "In the time of Charles the First [early 1600s], fishing with trained Great Cormorants was a regular sport in England" (Townsend, C. W., in Bent 1922, 236).

Breeding adult
Alain Richard

Juvenile
David Laliberte

China: white neck
Christian Artuso

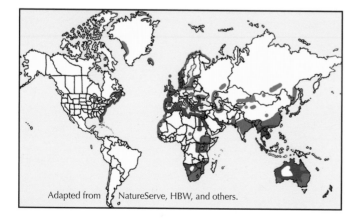
Adapted from NatureServe, HBW, and others.

NatureServe
Conservation Status

Global: .... Secure
Canada: ..... Apparently Secure
US: ....Critically Imperiled
Mexico: .... Not Applicable

www.natureserve.org

ROBERT ALVO

Red-faced Cormorant

Scientific: *Phalacrocorax urile*
Français: Cormoran à face rouge
México: Cormorán Cara Roja
Order: Suliformes
Family: Phalacrocoracidae

That this may be the least known of the bird species that regularly breed in North America should come as no surprise given its remote haunts, shy nature, and lack of value for human exploitation. Substantial gaps in knowledge exist for nearly every aspect of its life history.

The Red-faced Cormorant looks very much like its close relative, the Pelagic Cormorant, especially when the two are in nonbreeding plumages. They often co-occur and are often confused with each other. Many observations originally thought to have been made of Red-faced Cormorants actually involved both species and are therefore of limited use. On the other hand, many of the aspects of Red-faced Cormorant life history are often assumed, rightly or otherwise, to be similar to those of the Pelagic Cormorant. The Red-faced Cormorant isn't a very social cormorant and is generally a solitary feeder. Its colonies are usually small and dispersed,

John Hoyt

rarely consisting of more than 50 nests. Distribution and spacing are often affected by the availability of cliff ledges of the appropriate size (Causey 2002).

The Russian-American Company began introducing Arctic Foxes (*Vulpes lagopus*) to the Aleutian Islands in 1750 for fur farming, and later Red Foxes (*V. vulpes*). Through predation of eggs, chicks, and adults, foxes greatly reduced the breeding populations of all seabirds, including Red-faced Cormorants. Aleuts first expressed concern in 1811. Nevertheless, by the 1930s, over 450 islands had been stocked. Fox eradication programs were finally initiated in the 1980s, and Red-faced Cormorants have rebounded as a result. The US Fish and Wildlife Service continues fox eradication from most Aleutian islands to help all the seabirds breeding there (Bailey 1993).

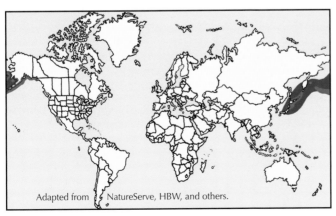

Adapted from NatureServe, HBW, and others.

NatureServe
Conservation Status

Global: .... Secure

Canada: .... Not Applicable

US: .... Secure

Mexico: .... Not Applicable

www.natureserve.org

Pelagic Cormorant

Scientific: *Phalacrocorax pelagicus*
Français: Cormoran pélagique
México: Cormorán Pelágico
Order: Suliformes
Family: Phalacrocoracidae

Even though gulls don't dive for food, the remains of deepwater prey have been found in their stomachs. How can this be and what does this have to do with cormorants? In numerous kinds of birds, parts of their prey that are difficult or impossible to digest, such as bones, teeth, feathers, fur, and the outer skeleton of some insects, are formed into pellets in the birds' gizzards. These pellets are regurgitated. Pellet-casters include owls, hawks, grouse, nightjars, swifts, kingfishers, shrikes, grebes, some thrushes, and cormorants.

In cormorants, pellets often contain many nematodes, suggesting that pellet-casting may aid in controlling internal parasites. Cormorants eject pellets throughout the year at the rate of almost one per night. During the nesting season, this is usually done in the darkness just before dawn when the bird is ready to leave the nest, possibly to avoid attracting gulls. Gulls fight each other for pellets, which they regard as food, and this activity disturbs the cormorant colony. Regurgitation just before daylight also allows a long time for the previous day's meal to be digested. Given that cormorants dive for their food, sometimes to great depths, gulls may thus be found with deepwater prey in their gut (Ainley et al. 1981). Even though the examination of pellets (and stomach samples) can be very useful when studying an animal's diet, the possibility that the remains came from another animal's deposited pellet or scat, or from a prey animal's gut, should be considered.

Despite the Pelagic Cormorant's four names listed above, it isn't pelagic, but rather it is an inshore species. It is rarely seen more than a few kilometers from land, taking mostly non-schooling fish over rocky substrates or kelp beds. Occasional sightings inland on rivers and lakes near marine coasts are thought likely to be a response to salmon runs (Hobson 2013).

Dennis Paulson

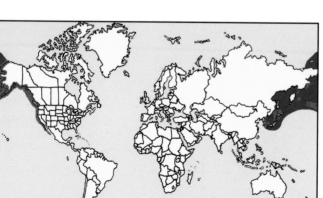

Adapted from NatureServe, HBW, and others.

NatureServe
Conservation Status

Global: .... Secure

Canada: ..... Apparently Secure

US: .... Secure

Mexico: .... Critically Imperiled

www.natureserve.org

Anhinga

Scientific: *Anhinga anhinga*
Français: Anhinga d'Amérique
México: Anhinga Americana
Order: Suliformes
Family: Anhingidae

The spread-wing posture is seen in large birds including New World vultures, pelicans, storks, cormorants, and anhingas when they stand with their wings open to the sides. Cormorants and anhingas use this posture to dry their wings after swimming, but anhingas do so also to absorb heat. The Anhinga is one of our most distinctive birds with its long snake-like neck and is often called "snake bird". Compared to cormorants, Anhingas have a more pointed bill that lacks the distinct hook at the tip. Also, Anhingas can kink their neck at a right angle. They are usually found in water or on perches (e.g., branches) above it. Their plumage is water-permeable, unlike that of most aquatic birds. This, along with their dense bones, allows Anhingas to float motionless at exactly the same density as water. They stalk fish and other prey while submerged usually less than 0.5 m with only their head and neck above water like a periscope. Fish are speared with a quick stab using both the upper and lower mandibles. Anhingas focus their efforts on slow-moving laterally-flattened fish. Backward-pointing serrations on the mandible tips keep the prey from sliding off the bill, but once up on

Giff Beaton

the surface the bird shakes the prey off the bill, tosses it in the air, catches it, and swallows it headfirst. The reason why Anhingas spend so much time in the spread-wing position is that their permeable plumage allows water to cool their skin. Anhingas thus play a thermoregulatory balancing act between time spent in water and time spent "hanging their wings to dry". They depend on warmth from the sun, which limits the northern extent of their distribution compared to that of cormorants (Frederick and Siegel-Causey 2000; Podulka et al. 2004). Observations after about 1990 suggest that the northern part of their distribution may be moving northward with climate change (eBird 2012).

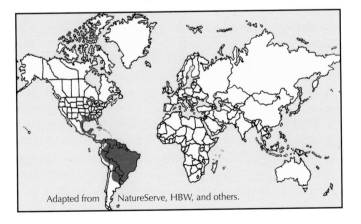

Adapted from NatureServe, HBW, and others.

NatureServe
Conservation Status

Global: .... Secure
Canada: .... Not Applicable
US: .... Secure
Mexico: .... Secure

www.natureserve.org

American White Pelican

Scientific: *Pelecanus erythrorhynchos*
Français: Pélican d'Amérique
México: Pelícano Blanco Americano
Order: Pelecaniformes
Family: Pelecanidae

Dixon L. Merritt wrote, "A wonderful bird is the pelican; his bill will hold more than his belican. He can take in his beak enough food for a week, but I'm damned if I see how the helican!" The American White Pelican forages on the surface, often cooperatively, using the bill pouch as a dip net. "Their favorite time for fishing on the seashore is during the incoming tide, as with it come the small fishes to feed upon the insects caught in the rise and upon the low forms of life in the drift as it washes shoreward. The larger fish follow in their wake, each from the smallest to the largest eagerly engaged in taking life in order to sustain life. All seabirds know this and the time of its coming well. The white pelicans that have been patiently waiting in line along the beach quietly move into the water and glide smoothly out so as not to frighten the life beneath. At a suitable distance from the shore they form into a line in accordance with the sinuosities of the beach, each facing shoreward and awaiting their leader's signal to start. When this is given all is commotion, the birds rapidly striking the water with their wings, throwing it high above them and plunging their heads in and out, fairly making the water foam as they move in an almost unbroken line, filling their pouches as they go" (Goss 1888, 26).

Juan Bahamon

The American White Pelican's global population was at risk until the 1960s because of changing water levels, human disturbance, and possibly contaminants, but it has recovered. It remains Apparently Secure because of its high vulnerability to disturbance, uncertain habitat protection, and diseases (NatureServe 2014b).

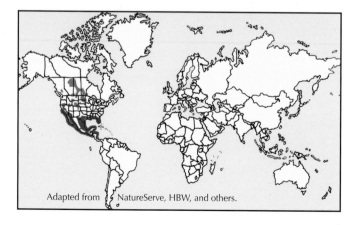

Adapted from NatureServe, HBW, and others.

NatureServe
Conservation Status

Global: ...... Apparently Secure
Canada: .... Vulnerable
US: .... Apparently Secure
Mexico: ..... Apparently Secure

www.natureserve.org

Brown Pelican

Scientific: *Pelecanus occidentalis*
Français: Pélican brun
México: Pelícano Café
Order: Pelecaniformes
Family: Pelecanidae

At the age of 3–4 weeks the young Brown Pelican is large enough to swallow whole fish, which it takes by reaching its bill into the parent's throat, thus making it disgorge. As in some other birds, the earlier-hatched young often kills its siblings, either directly by pecking at them or indirectly by preventing them from feeding or forcing them out of the nest. This is called siblicide.

Brown Pelicans almost became extirpated from North America between the late 1950s and early 1970s due to two organochlorine pesticides. Endrin killed the birds directly, whereas DDT caused eggs to have thin shells and thus to break easily. Recovery of the species after a great reduction in the use of these pesticides has been impressive (Shields 1992).

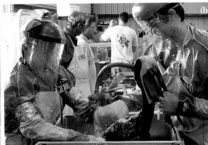

Sequence of photos showing cleaning process for birds oiled during the 2010 Deepwater Horizon disaster.

(a) Oiled birds.

(b) Wash.

(c) Rinse cycle.

(d) Clean, awaiting freedom.

Nonbreeding adult shown in (e).

PHOTOGRAPHERS:
Sarah Hechtenthal (a–d)
Edgar T. Jones (e)

See Magnificent Frigatebird account, p.98, for photos of release after cleaning.

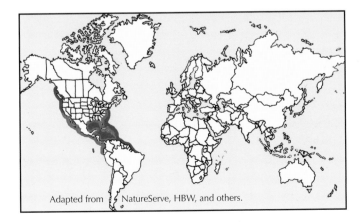

Adapted from NatureServe, HBW, and others.

NatureServe
Conservation Status

Global:...... Apparently Secure

Canada:.......... Not Applicable

US:............ Apparently Secure

Mexico:..... Apparently Secure

www.natureserve.org

American Bittern

Scientific: ***Botaurus lentiginosus***
Français: Butor d'Amérique
México: Avetoro Norteño
Order: Pelecaniformes
Family: Ardeidae

If you spent all your time in a marsh and wanted to communicate over a long distance with someone whom you couldn't see through the dense vegetation, you would likely find it easier to do so by using low-frequency sounds rather than high-frequency ones. Indeed the American Bittern is a large marsh dweller that vocalizes to mates and other individuals during the breeding season using low, resonant calls described as *pump-er-lunk* or *dunk-a-doo* (Lowther et al. 2009). "These notes have been likened to the sound made by an old wooden pump in action.... The bittern is not an active bird. It spends most of its time standing under cover of vegetation, watching and waiting for its prey or walking slowly about in its marsh retreat, raising each foot slowly and replacing it carefully; its movements are stealthy and noiseless, sometimes imperceptibly slow so as not to alarm the timid creatures that it hunts. When standing in the open or when it thinks it is observed, it stands in its favorite pose with its bill pointed upward and with its body so contracted that its resemblance to an old stake is very striking" (Bent 1926, 79–80).

Robert E. Gehlert

Christian Artuso

Steve Zamek

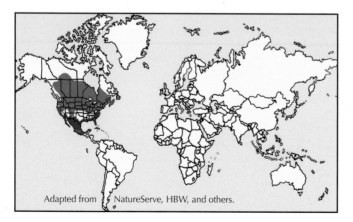
Adapted from NatureServe, HBW, and others.

NatureServe
Conservation Status

Global: ...... Apparently Secure

Canada: .... Secure

US: .... Apparently Secure

Mexico: ...Presumed Extirpated

www.natureserve.org

Least Bittern

Scientific: *Ixobrychus exilis*
Français: Petit Blongios
México: Avetoro Menor
Order: Pelecaniformes
Family: Ardeidae

The Least Bittern is strikingly smaller than the American Bittern. Its diminutive size, highly secretive nature, and rarity in many areas make it very difficult to find even when it is known to be present. Like the American Bittern, it freezes with its bill pointing up when alarmed. It lives in marshes that have dense, tall, emergent vegetation dotted with areas of open water.

This species has declined greatly in abundance in North America due to marsh drainage. Yet its fairly large global range, common status in parts of its range, and its use of artificial marshes combine to make it globally Secure.

To maintain its status in North America, the marshes that remain must be conserved or improved, especially those larger than five hectares. Its habitat also needs to be protected from silt, eutrophication, and chemicals. Managed wetlands should maintain appropriate ratios of vegetated marsh area to open water area. Managing wetlands as complexes with different water-level regimes is useful for ensuring that manipulations of parts of the complex (e.g., for waterfowl) allow some suitable habitat for this species at all times. Nutrient-poor or acidic wetlands can be improved by liming and fertilizing dykes and surrounding fields. Exotic plants can be controlled using herbicides, physical removal, and burning. Proper management requires standardized techniques for monitoring its numbers and breeding success and to allow comparisons to be made between years and between sites (NatureServe 2014b). In addition, such management is likely to enhance the environment not only for Least Bitterns, but also for other species of plants and animals that require high-quality wetland habitat.

Gord Belyea

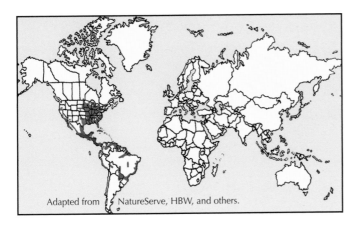

Adapted from NatureServe, HBW, and others.

NatureServe
Conservation Status

Global: .... Secure

Canada: ..... Apparently Secure

US: .... Secure

Mexico: .... Secure

www.natureserve.org

Great Blue Heron

Scientific: ***Ardea herodias***
Français: Grand Héron
México: Garza Morena
Order: Pelecaniformes
Family: Ardeidae

Recently-fledged herons often disperse hundreds of kilometers in all directions to find new sources of food, thus avoiding crowding near colonies. This behavior may also precede establishment of new colonies. In this post-fledging dispersal, individuals that travel north of their breeding range (in the Northern Hemisphere) take advantage of habitats that were unavailable (e.g., frozen) to their parents in spring. Later in the season these young birds join their species' main southward migration. Whereas our simplified global range maps don't illustrate migration or dispersal ranges, maps in many regional field guides do so and should be consulted when this information is required (e.g., Dunn and Alderfer 2011).

In the northern parts of their range, adult Great Blue Herons flock on the ground for several days in spring before entering their colonies, which are usually located high up in trees, but the function of these "gathering grounds" is unknown (Vennesland and Butler 2011).

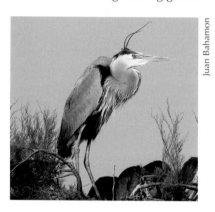

Juan Bahamon

Juan Bahamon

Adapted from NatureServe, HBW, and others.

NatureServe
Conservation Status

Global: .... Secure

Canada: .... Secure

US: .... Secure

Mexico: .... Secure

www.natureserve.org

Great Egret

Scientific: *Ardea alba*
Français: Grande Aigrette
México: Garza Blanca
Order: Pelecaniformes
Family: Ardeidae

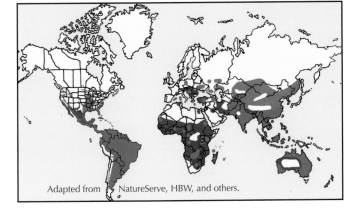

How can it be that the parents of some birds sit idly at the nest while one of their offspring stabs a sibling to death with its bill? Presumably the inaction of the parents and the actions of the dominant offspring are adaptive, but how so? If the amount of food available for the young is limited such that only a portion of the nestlings can fledge, then perhaps this makes sense.

Great Egrets lay up to six eggs, but the North American average is only about three. Eggs are laid at intervals of one or two days, and incubation starts upon laying of the first egg, so hatching occurs asynchronously. This allows earlier-hatched nestlings to grow larger than their egg-bound siblings. In years when food is plentiful, all the chicks in the nest can be fed and they may all survive. However, if food is limited the large chicks may kill the small ones directly, or push them out of the nest and allow them to starve. A small number of chicks having a high probability of survival to fledging is better than more chicks having little or no chance (Podulka et al. 2004).

Birds have many types of sleeping places (Skutch 1989), and finding them can be even more challenging for field biologists than finding nests. Great Egrets often gather in large numbers in trees just before dark, spend the night there, then scatter in the morning (Bent 1926).

Juan Bahamon

Juan Bahamon

Cynthia Pekarik

Adapted from NatureServe, HBW, and others.

NatureServe
Conservation Status

Global: .... Secure
Canada: .... Vulnerable
US: .... Secure
Mexico: .... Secure

www.natureserve.org

Snowy Egret

Scientific: *Egretta thula*
Français: Aigrette neigeuse
México: Garza Dedos Dorados
Order: Pelecaniformes
Family: Ardeidae

This species has a greater repertoire of foraging techniques than does any other heron, but it spends much time in chasing and slow pursuit behaviors, which take considerable energy. Thus, the Snowy Egret must spend more time foraging than do related species (Parsons and Master 2000).

The Snowy Egret was slaughtered in great numbers in the late 1800s because of the huge demand for its beautiful plumes, which commanded twice the price of gold. "There, strewn on the floating water weed and also on adjacent logs were at least 50 carcasses…, nearly one-third of the rookery…, the birds having been shot off their nests containing young.… Parentless young ... were seen trying in vain to attract the attention of passing egrets that were flying with food in their bills to feed their own young, and it was a pitiful sight to see these starvlings [sic] with outstretched necks and gaping bills imploring the passing birds to feed them" (Mattingley, A. H. E., in Bent 1926, 153). This madness ended after signing of the Migratory Bird Convention in 1916.

Carol Blackard

Juan Bahamon

This bird has a caterpillar on its left leg.

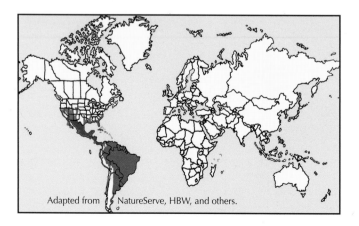

Adapted from NatureServe, HBW, and others.

NatureServe
Conservation Status

Global: .... Secure

Canada: .... Critically Imperiled

US: .... Secure

Mexico: .... Secure

www.natureserve.org

Little Blue Heron

Scientific: *Egretta caerulea*
Français: Aigrette bleue
México: Garza Azul
Order: Pelecaniformes
Family: Ardeidae

Kevin Wallace

"There is, however, a human enemy, unconscious perhaps of his evil deeds, who causes considerable havoc whenever he indulges in his supposedly harmless sport in a heron rookery; and that is the bird photographer who sets up his blind in a rookery and keeps herons off their nests, often for long periods.... The crows and vultures had cleaned out practically all the nests anywhere near our blinds…, and the ground was strewn with broken eggshells all over the rookery" (Bent 1926, 175).

The past two decades have seen a surge in bird photography perhaps more impressive than the concomitant increase in North Americans who bird. In addition, many birdwatchers themselves have become bird photographers. This often puts more pressure on the birds, which may react negatively to human presence at any time of year. Most birding clubs have their own code of ethics or recommend one for their members to follow. The same codes generally apply to nature photography. The American Birding Association's code, for example, has four main principles: "1) Promote the welfare of birds and their environment; 2) Respect the law, and the rights of others; 3) Ensure that feeders, nest structures, and other artificial bird environments are safe; and, 4) Group birding, whether organized or impromptu, requires special care". For details, see www.aba.org/about/ethics.html/.

The Little Blue Heron is the only Ardeidae species with distinct color morphs for juveniles (i.e., first-year white birds) and adults (blue). Immatures become patchy white-and-blue over the second year. In contrast, the Reddish Egret has two adult color morphs: dark and white. In southern Florida, one can practice distinguishing white birds of six breeding Ardeidae species. This does not include albinos, which are rare in most life forms in which albinism frequency in the wild has been gauged.

Juan Bahamon

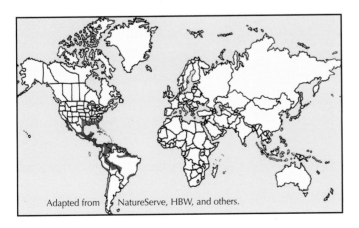

Adapted from NatureServe, HBW, and others.

NatureServe
Conservation Status

Global: .... Secure
Canada: ..... Critically Imperiled
US: .... Secure
Mexico: .... Secure

www.natureserve.org

Tricolored Heron

Scientific: *Egretta tricolor*
Français: Aigrette tricolore
México: Garza Tricolor
Order: Pelecaniformes
Family: Ardeidae

"How agile and graceful they were as they darted about in pursuit of their prey. With what elegance and yet with what precision every movement was made. For harmony in colors and for grace in motion this little heron has few rivals.... I shall never forget my first impression of this elegant 'lady of the waters'" (Bent 1926, 168). Part of the Tricolored Heron's charm is certainly its very dainty physique.

Colonies are usually located in landscapes with different wetland types, and this heron takes advantage of them by using various foraging methods, preferring areas with some open water and some low vegetation. It may feed alone or in groups, sometimes with other species. Being mostly a surface feeder, its foraging success is increased by low water-oxygen content and other conditions that force prey to the surface. It may use Double-crested Cormorants and Pied-billed Grebes as "beaters", that is as mechanisms for disturbing prey to make it more accessible.

The Tricolored Heron has a very interesting relationship with American Alligators (*Alligator mississippiensis*). On one hand it must be wary of them when it is foraging in water or resting on shore. Alligators will also take chicks that fall out of nests, and may hunt young birds that are perched on lower branches. These massive reptiles can even jump or climb up to two meters to seize nestlings. On the other hand, alligators often keep mammalian predators such as raccoons away from colonies—indeed, Tricolored Herons seem to prefer nesting above alligators.

Richard Higgins

The decline in Tricolored Heron numbers in the famous Florida Everglades is thought to be the result of the reduced volume of fresh water flowing into that estuary because this species is dependent on productive estuaries. In Latin America, Tricolored Heron habitat (estuaries and marine swamps) is being lost to commercial shrimp aquaculture, coastal development, the rise in sea level, and the increased frequency of hurricanes (Frederick 2013). Nevertheless, it is still globally Secure.

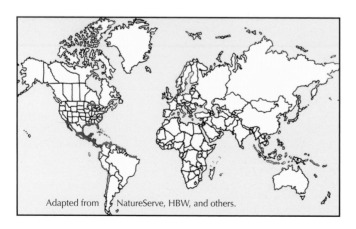

Adapted from NatureServe, HBW, and others.

NatureServe

Conservation Status

Global: .... Secure
Canada: .... Not Applicable
US: .... Secure
Mexico: .... Secure

www.natureserve.org

Reddish Egret

Scientific: *Egretta rufescens*
Français: Aigrette roussâtre
México: Garza Rojiza
Order: Pelecaniformes
Family: Ardeidae

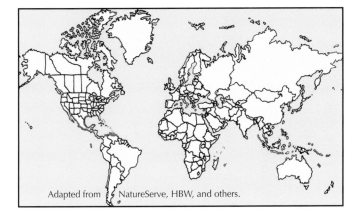

Oh Todd, you're SOOO handsome !!!!!

The Reddish Egret is dimorphic, meaning that it has two color morphs: a dark one and a white one. The latter was originally thought to be a juvenal plumage (see Little Blue Heron account, p. 115) until mixed pairs were found breeding successfully in the 1870s.

Crest-raising is often used in aggressive and courtship behavior. "The plumes of the head, neck, breast, and back stand out like the quills of a porcupine, giving the bird quite a formidable appearance, terrifying to its enemies perhaps, but probably pleasing to its mate" (Bent 1926, 158). This, along with its erratic and comical foraging techniques, which include spinning, hopping, dashing, jumping, and various open-winged behaviors to fool fish, are this species' most distinguishing characteristics in comparison to other herons, which are generally less animated (Lowther and Paul 2002).

The Reddish Egret is our least known and rarest heron, having been almost extirpated from the US by plume hunting before 1900. The US population recovered somewhat and is now estimated at 900–950 pairs in Texas, 350–400 in Florida, 60–70 in Louisiana, and 5–10 in Alabama. Mississippi's short coastal strip has none (Vermillion and Wilson 2009). The major threat to this coastal species is habitat loss from coastal development, dredging, "bulkheading" (the creation of retaining walls), changes in water levels, cattle-grazing on some islands, and the harvest of mangroves for charcoal (Lowther and Paul 2002).

Juan Bahamon

Adults: white morph and dark morph

Juvenile white morph

Adapted from NatureServe, HBW, and others.

NatureServe
Conservation Status

Global: ...... Apparently Secure

Canada: .... Not Applicable

US: .... Apparently Secure

Mexico: ..... Apparently Secure

www.natureserve.org

Cattle Egret

Scientific: ***Bubulcus ibis***
Français: Héron garde-boeufs
México: Garza Ganadera
Order: Pelecaniformes
Family: Ardeidae

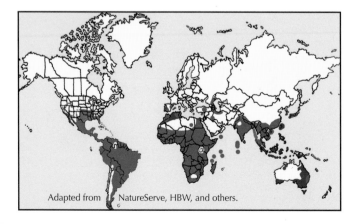

This is the only bird that has both crossed to the New World from the Old World without direct help from humans, and phenomenally expanded its numbers and breeding range. Eurasian Wigeons and Lesser Black-backed Gulls, Old World birds found regularly in much of North America during their breeding season, aren't known to breed here regularly.

Conversely, Rock Pigeons, European Starlings, and House Sparrows, now occurring throughout much of the New World, were introduced willfully by humans. Other species, introduced for hunting, haven't expanded as much. Thus, given that it hasn't been introduced here and that it breeds regularly, the Cattle Egret is included in the main part of this book while the above-mentioned five species are relegated to the Appendix.

Cattle Egrets probably crossed the Atlantic Ocean from Africa to South America in the late 1800s, likely using the shortest route, and arrived in North America in the early 1940s. They continue to move northward. Renowned wanderers, they couldn't colonize the New World until cattle ranching became widespread. Their ability to forage along marine shorelines (e.g., on exhausted small migrant birds) also helped them to disperse (Telfair 2006).

Studying the Cattle Egret is easy and fascinating. Many papers have been published on it in many languages. In its original African foraging habitat of short-grass meadow, it followed moving African Buffalos (*Syncerus caffer*), which stir up insects. The Cattle Egret now uses many

Larry Master

hosts, even ostriches and tortoises. It often perches on the host's back or whips sedentary ones into action by making restless flights nearby. In Kenya, and perhaps elsewhere, it selects hosts that take 5–15 steps/min so it can eat the energetically optimal number of food items. It also uses cooperative "leapfrog feeding", in which the first birds to land near a host are overflown by following birds that in turn are overflown by others—each group stirs up insects for the next birds. Leapfrogging occurs mostly at farm tractors, which also stir up insects (Telfair 2006). Perhaps Cattle Egrets try to start tractors in disuse with their restless flights.

Adapted from NatureServe, HBW, and others.

NatureServe
Conservation Status

Global: .... Secure
Canada: .... Critically Imperiled
US: .... Secure
Mexico: .... Secure

www.natureserve.org

Green Heron

Scientific: ***Butorides virescens***
Français: Héron vert
México: Garcita Verde
Order: Pelecaniformes
Family: Ardeidae

"The nest itself is a simple affair,… ill adapted, it would seem, to hold eggs when the tree branches wave in the wind, for it is a flat platform of sticks, destitute of any sort of lining and not cup shaped…. The nest is so thin and flimsy that one can sometimes look through it from below and see the eggs" (Townsend, C. W., in Bent 1926, 187). Why are some Green Heron nests so flimsy? One possibility is that such nests are less noticeable to predators. I once found a nest that looked like a construction from a previous year. Hearing birds above, I continued to search the tree canopy and eventually found not only that six loosely-built nests were present, but that most had Green Herons on them or nearby. Eggs would be less likely to fall through elaborate, strongly constructed nests during incubation, but on the other hand losing an egg here and there through the Green Heron's flimsy nests may be an acceptable trade-off against the benefits of eluding predators. Parents can rapidly fix nests that lose some sticks while the young hang onto other branches (Davis and Kushlan 1994).

Dan Parent

Juan Bahamon

NatureServe
Conservation Status

Global: .... Secure
Canada: ..... Apparently Secure
US: .... Secure
Mexico: .... Secure

www.natureserve.org

Adapted from NatureServe, HBW, and others.

Black-crowned Night-Heron

Scientific: *Nycticorax nycticorax*
Français: Bihoreau gris
México: Garza Nocturna Corona Negra
Order: Pelecaniformes
Family: Ardeidae

Apart from doing so when they feel nauseous, birds may disgorge their food when provoked by parasites such as jaegers. Some birds, such as herons, forage away from their colonies and disgorge food to their young in the nest. Young herons themselves may disgorge food when they feel threatened. For example, Black-crowned Night-Heron nestlings are aggressive with intruders, and disgorge on them, be they predators or researchers. By reducing the weight of nestlings, this may allow for a quicker escape. This behavior may also serve as a food offering that distracts the intruder away from the young bird.

This species is used to evaluate contaminant levels throughout the world. Its position high up the food chain allows contaminants levels to be representative of the entire food chain. Another reason for its usefulness in this regard is its cosmopolitan distribution. Because it is the world's most widespread heron, meaningful comparisons of contaminant levels can be made among different parts of the world without having to take into account differences between species. Also, this species' colonial nature makes it possible to sample a relatively large number of individuals, pairs, or families within a small area, thus reducing research costs (Hothem et al. 2010).

Christian Artuso

Cynthia Pekarik

Dan Parent

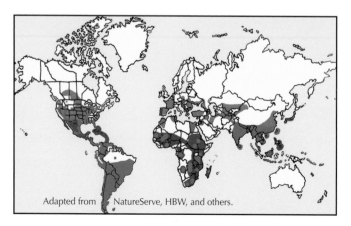
Adapted from NatureServe, HBW, and others.

NatureServe
Conservation Status

Global: .... Secure

Canada: ..... Apparently Secure

US: .... Secure

Mexico: .... Secure

www.natureserve.org

Yellow-crowned Night-Heron

Scientific: *Nyctanassa violacea*
Français: Bihoreau violacé
México: Garza Nocturna Corona Clara
Order: Pelecaniformes
Family: Ardeidae

The genus name *Nyctanassa* means "night queen" and refers to this species' beauty and its habit of foraging at night. Poor lighting characterizes most of its habitats, which include forested wetlands, swamps, and bayous. The Yellow-crowned Night-Heron is a crustacean specialist, taking mostly crabs and crayfish. Not surprisingly, much of its life history is related to this predilection. Along the Atlantic Coast, the initiation of breeding in spring coincides with the timing of crab emergence from the ocean, which itself depends on environmental temperatures. Because temperature patterns vary among years at a given location and also with latitude, the timing of the onset of breeding can vary considerably. Northern night-heron populations migrate south in the fall to forage on crabs that are active all year. The Yellow-crowned Night-Heron is the most sedentary forager of the ten species of North American herons and egrets, spending at least 80% of its time standing, head-swaying, neck-swaying, or pecking. Bill size is related to the size of crabs that it eats. For example, two subspecies that occur only on certain islands and that feed on armored land crabs have heavier and wider bills than subspecies with larger distributions.

Dennis Paulson

Juan Bahamon

Whereas Black-crowned Night-Herons avoid nesting near human habitation, Yellow-crowned often nest in wooded park-like neighborhoods that have sparse or absent understories. Such colonies may make up a substantial proportion of the regional population, with many pairs nesting over roofs, driveways, and roads. This often leads to conflicts with humans (Watts 2011).

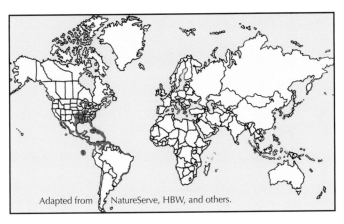
Adapted from NatureServe, HBW, and others.

NatureServe
Conservation Status

Global: .... Secure

Canada: .... Not Applicable

US: .... Secure

Mexico: .... Secure

www.natureserve.org

White Ibis

Scientific: *Eudocimus albus*
Français: Ibis blanc
México: Ibis Blanco
Order: Pelecaniformes
Family: Threskiornithidae

The White Ibis is a nomad. Apart from migrating and undertaking significant postbreeding dispersals as do other herons, its populations often move regionally before the breeding season. Rainfall causes populations to shift between the coast and the interior, and colony sites are regularly changed. White Ibises may be stimulated to nest in a colony of earlier-nesting related species. Their nomadic behavior and flexible timing of breeding onset allow them to be at the right place at the right time as their shallow-water foraging sites change with climatic conditions.

White Ibises are very gregarious. They nest in large, dense colonies, and the young form crèches when leaving nests. Foraging almost always involves flocks. Roosts are communal, and the birds fly to and from roosts, colonies, and feeding sites in flocks. Roost sites may generate breeding colonies when daytime bachelor parties form there and on the surrounding ground. Nest building, often not synchronous in a large colony, is usually completed in four days in "neighborhoods" of 25–50 pairs. Nonbreeding second-year birds often act as "helpers" by moving among nests to preen and shade nestlings. Nestlings beg from helpers, but feeding has not been reported. Helpers work only when adults are absent, and are sometimes chased from the site by returning parents. Perhaps the helpers are practicing for their own future benefit.

Juan Bahamon

To maintain their salt balance, nestlings need food from freshwater wetlands rather than from the sea. Even parents nesting on coastal islands forage for their young inland—in a 10,000-pair colony that had no birds the year after a hurricane, and in which the colony site was not affected, the nearby foraging sites were flooded with seawater. As nestlings start fledging, parents shift their foraging from freshwater habitats (e.g., cypress swamps) to saltwater ones (e.g., salt marshes) (Heath et al. 2009).

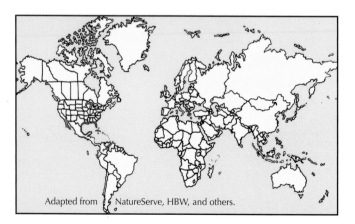

Adapted from NatureServe, HBW, and others.

NatureServe
Conservation Status

Global: .... Secure

Canada: .... Not Applicable

US: .... Secure

Mexico: .... Secure

www.natureserve.org

Glossy Ibis

Scientific: *Plegadis falcinellus*
Français: Ibis falcinelle
México: Ibis Cara Oscura
Order: Pelecaniformes
Family: Threskiornithidae

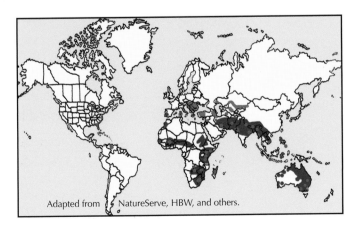

The Glossy Ibis's nomadic habits and its widespread postbreeding dispersals in all directions may have prevented it from differentiating into recognized subspecies. It colonized North America from Europe in the 1800s and remarkably increased its distribution from the 1950s through the 1970s. However, its numbers later declined in many areas (Davis and Kricher 2000).

Crayfish is a highly regarded food for humans living in the southern US, with Louisiana producing the greatest harvest of this crustacean in North America in both aquaculture and wild capture fisheries. Commercial crayfish aquaculture has blossomed in Louisiana in the past few decades, and likely caused the observed increase in populations of colonial wading bird species, especially ones that eat many crayfish, such as the Glossy Ibis. Commercial crayfish ponds provide a rich and predictable foraging habitat (Fleury and Sherry 1995).

Larry Master

This ibis is thought to be a "core species" in that its presence attracts other closely related wading species to forage in the same area. For example, it seems to have a "beater-follower" relationship with the Snowy Egret in that, as the ibis probes for prey and stirs it up in the shallow water, the egret takes advantage of the more readily available prey. Boat-tailed Grackles steal crayfish from Glossy Ibises and chase them up to 30 m above the ground when they try to fly away.

The Glossy Ibis usually nests in mixed-species colonies with other wading species. Within a colony it tends to nest in denser vegetation than other species do and often its nests are dispersed throughout the colony (Davis and Kricher 2000).

Adapted from NatureServe, HBW, and others.

NatureServe
Conservation Status

Global: .... Secure
Canada: .... Not Applicable
US: .... Apparently Secure
Mexico: .... Secure

www.natureserve.org

White-faced Ibis

Scientific: *Plegadis chihi*
Français: Ibis à face blanche
México: Ibis Ojos Rojos
Order: Pelecaniformes
Family: Threskiornithidae

Even though the White-faced Ibis has two widely separate populations, one in North America and the other in southern South America, no subspecies are recognized. Rather, it was previously considered a subspecies of the Glossy Ibis. Northern North American subpopulations migrate south annually, whereas most South American subpopulations do not migrate northward. Nevertheless, as in other large wading species, juvenile birds disperse widely in both South and North America after the breeding season, although apparently not widely enough for the two global populations to mix (Ryder 1998). The White-faced Ibis is abundant in parts of its South American range, where it is the most common of the region's six ibis species (de la Pena and Rumboll 1998), and where one million pairs were estimated to occur in 2006. In contrast, only 150,000 birds are estimated for North America even though the species' range has expanded westward since the 1980s (Matheu et al. 2014). The population had declined in the 1960s and 1970s because of habitat loss and the use of DDT, but has since been increasing rapidly.

Netta Smith

This species feeds in large flocks of up to 1000 or more birds, using mostly freshwater marshes, ponds, rice fields, flooded pastures, irrigated crop fields, and the edges of large bodies of water. It sometimes feeds in brackish waters. In the Great Basin of the US, the loss of wetland feeding habitat to human use is partly replaced by irrigated crop fields, of which those with alfalfa are greatly preferred to those of corn, wheat, barley, and oats. Fields are usually abandoned once the pools dry up, but often not before the birds have trampled the field enough to compact the soil and/or flatten the alfalfa, making it impossible to cut for hay (Ryder and Manry 1994).

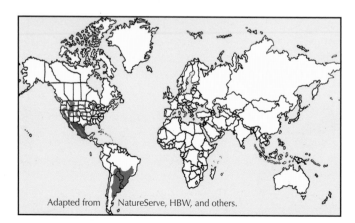

Adapted from NatureServe, HBW, and others.

NatureServe
Conservation Status

Global: .... Secure

Canada: ..... Critically Imperiled

US: .... Apparently Secure

Mexico: .... Secure

www.natureserve.org

Roseate Spoonbill

Scientific: *Platalea ajaja*
Français: Spatule rosée
México: Espátula Rosada
Order: Pelecaniformes
Family: Threskiornithidae

"All [Man's] sordid mind can grasp is the thought of a pair of pretty wings and the money they will bring when made into ladies' fans!" (Bent 1926, 13). During the late 1800s, this species was decimated by collecting for the feather trade and from disturbance of nesting colonies for the plumes of other species (Dumas 2000).

Many societies throughout the world have used the feathers and skins of birds for warmth and as ornaments, but in the late 1800s in the US, plumes were so popular for women's hats that egret and heron populations plummeted. As is often the case with disasters, good came out of this free-for-all in the southern US swamps in the form of a conservation movement. J. J. Audubon was already concerned about bird declines in the 1840s. By the 1870s, Passenger Pigeons had all been harvested and it was now the Eskimo Curlew's turn. The American Ornithologists' Union was founded in 1883, and by 1887 over 300 local Audubon chapters started protecting birds.

Juan Bahamon

Thoreau, Emerson, and Whitman wrote about the importance of nature, followed by Burroughs and Muir. In 1887, the Smith College Audubon Society was created by Fannie Hardy and Florence Merriam to persuade women to stop wearing feathers. The federal Lacey Act of 1900 banned interstate shipments of birds killed illegally, and the country's first wildlife warden patrolled the western Everglades. He and two other wardens were murdered by plume hunters. The Audubon Plumage Act of 1910 ended widespread trade in bird plumes from US birds, while the 1911 Bayne Act prohibited the sale of wild game in markets and restaurants. In 1913, plume imports were banned. Bird protection in the US and Canada was initiated by the Migratory Bird Convention as both countries adopted acts to protect birds, and this stimulated natural land protection in both countries (Podulka et al. 2004).

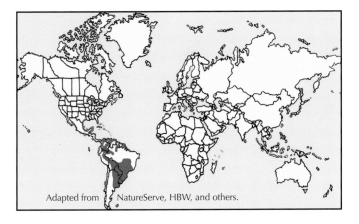

Adapted from NatureServe, HBW, and others.

NatureServe
Conservation Status

Global: .... Secure

Canada: .... Not Applicable

US: .... Apparently Secure

Mexico: .... Secure

www.natureserve.org

Black Vulture

Scientific: *Coragyps atratus*
Français: Urubu noir
México: Zopilote Común
Order: Accipitriformes
Family: Cathartidae

New World (NW) and Old World (OW) vultures are an excellent example of convergent evolution. Many of their similarities evolved separately, including their bare necks and heads, hooked bills, huge wings used for soaring, scavenging foraging behavior, and the common co-existence of several species. In a given region, there are often three size groups. The largest species, such as NW condors and OW Griffon Vultures (*Gyps fulvus*), are dark with a white neck ruff, travel great distances while foraging, visit carcasses in large numbers, and use their long necks to reach inside carrion. Medium sized species, such as NW King Vultures (*Sarcoramphus papa*) and OW Lappet-faced Vultures (*Torgos tracheliotos*), are more brightly colored, may have smaller foraging ranges, do not occur in large numbers at carcasses, have short necks, and eat skin and tough tissues. The "small" species, such as the NW Black Vulture and the OW Hooded Vulture (*Necrosyrtes monachus*), travel far while foraging and are food generalists, not being as dependent on large mammal carcasses. At large carcasses they feed mostly on food left by the other species on the ground or on the bones. Nevertheless, differences in behavior and detailed anatomy clearly show that the NW and OW vultures are very distant relatives, as Thomas Huxley, "Darwin's Bulldog", alluded to in 1876.

Juan Bahamon

The Black Vulture's scientific name means, "black as a raven and dressed in mourning". This species exploits the Turkey Vulture's strong sense of smell by soaring higher for a large field of view and following it to carcasses. Black Vultures also keep their eyes on each other in case one has found a carcass. Once many of them have arrived they displace the Turkey Vultures.

The Black Vulture has been closely associated with humans in rural Latin America, likely allowing its numbers to expand greatly in the last 400 years. It is the most abundant NW vulture (Buckley 1999, Houston et al. 2013).

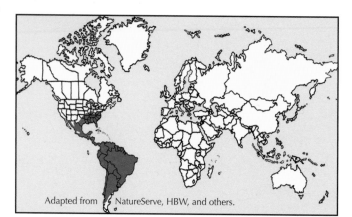

Adapted from NatureServe, HBW, and others.

NatureServe
Conservation Status

Global: .... Secure

Canada: .... Not Applicable

US: .... Secure

Mexico: .... Secure

www.natureserve.org

Turkey Vulture

Scientific: ***Cathartes aura***
Français: Urubu à tête rouge
México: Zopilote Aura
Order: Accipitriformes
Family: Cathartidae

Vultures play a key role in nature by disposing of carrion, which otherwise would be breeding grounds for disease. The very strong acid in their digestive system kills viruses and bacteria in carrion, such that vulture droppings and dry regurgitated pellets are clean and free of disease. In fact, the Turkey Vulture's genus name *Cathartes* means to clean, referring to the cleaning it does of animal bones. Turkey Vultures often start feeding on a carcass by extracting an eye and eating it.

Turkey Vulture researcher C. Stuart Houston observes that "the Turkey Vulture is smelly and ugly, but beautiful in soaring flight. It soars all the way to Venezuela from Saskatchewan, eating very little food en route", as his wing tagging program of 1100 individuals to date has shown. In flight, it raises its wings to an angle above horizontal making a shallow "V" (for "Vulture"), with the separated tips of the primary feathers often curled higher.

Teetering from side to side like a tight-rope walker, it can smell and find invisible carcasses below the forest canopy.

This species is the New World's most widely distributed vulture. In the US it causes more bird strike damage and human fatalities in military aircraft than does any other species (Kirk and Mossman 1998).

PHOTOGRAPHERS:
(a) Brent Terry
(b) Dennis Paulson
(c) Brent Terry
(d) Brent Terry

Playing dead with two eggs

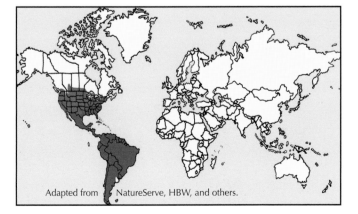
Adapted from NatureServe, HBW, and others.

NatureServe
Conservation Status

Global: .... Secure
Canada: .... Secure
US: .... Secure
Mexico: .... Secure

www.natureserve.org

California Condor

Scientific: *Gymnogyps californianus*
Français: Condor de Californie
México: Cóndor Californiano
Order: Accipitriformes
Family: Cathartidae

As of 2014, there are just over 400 California Condors in the world, with about half of them in captivity and the other half in the wild in California, Arizona, Utah, and Baja California. The species was apparently rare and declining in the late 1800s largely as a result of ingesting lead shot and bullet fragments in carrion. Condor numbers declined through the 1900s, and the species' range became restricted to southern California by the 1950s. Population size was an estimated 150 birds in 1950, 60 in 1968, and 30 in 1978.

In 1982, at an all-time low of 21 birds, a bold and controversial decision was made to try saving the species by taking the entire population into captivity along with eggs from wild pairs. By 1987, all 27 birds were in captivity. Breeding there succeeded in 1988, and captive production of birds since then has been successful. Releases to the wild began in 1992. Wild birds first laid eggs in 2001, but were unsuccessful. Wild fledglings were finally produced in 2002, but viable wild populations have not been established to date because of continued high mortality from lead poisoning. The species is still dependent on releases of captive-reared birds and on intensive management: wild condors are regularly trapped, tested, and treated for lead. Several individuals have been saved more than once from death by lead poisoning. This expensive management depends on the wild birds being trained to depend on lead-free carcasses provided by humans at feeding stations so that they can easily be trapped and treated. As the population grows and the birds feed farther away, they again feed on contaminated carrion (Snyder and Schmitt 2002; Wells 2007; NatureServe 2014b).

Ultimately, bans on the use of lead shot are needed, not just to save the California Condor from extinction, but also to protect other biota. In 2013, California became the first US state to ban lead shot. Other jurisdictions are expected to follow suit.

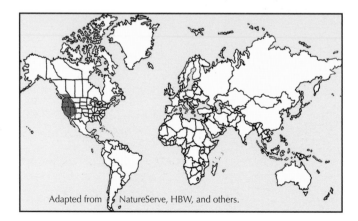

Adapted from NatureServe, HBW, and others.

NatureServe
Conservation Status

Global: ..... Critically Imperiled

Canada: .... Not Applicable

US: .... Critically Imperiled

Mexico: .... Critically Imperiled

www.natureserve.org

Osprey

Scientific: *Pandion haliaetus*
Français: Balbuzard pêcheur
México: Águila Pescadora
Order: Accipitriformes
Family: Pandionidae

This is North America's only raptor that eats live fish almost exclusively. It plunges feet-first into the water, reaching down to one meter below the surface. The easiest fish for it to catch there are bottom-feeders, which may be searching downward for food rather than upward for predators. The next-easiest prey are the non-fish-eating fish swimming in the top meter of deeper waters; they are slower than fish-eating fish. Ospreys thus favor foraging areas with abundant shallow clear waters, such as human-made reservoirs. An Osprey's catch may weigh 10–50% as much as the bird, whose wing shape and bone structure may be key to allowing it to emerge and take off while wet. The Osprey then maneuvers the fish in its feet to aim it forward, thus reducing aerodynamic drag.

Artificial nesting structures, more than reservoirs, have perhaps helped Ospreys increase their breeding distribution and numbers after the contaminant-induced declines of the 1950s–1970s.

Juan Bahamon

While some Ospreys continue to nest on trees, cliffs, large shoreline boulders, and ground sites on predator-free islands, most now use artificial structures such as communication towers, utility lines, nautical channel markers, and nesting poles erected for this species (often with platforms at the top). Causes of this shift in nest sites are: increased predation by Raccoons (*Procyon lotor*), whose populations have increased; the loss of tall trees; and, shoreline development. Nests on artificial structures have high success rates because they are usually stable.

Male Ospreys may defend two nests that are close together to breed with two females concurrently. Failed breeding Ospreys may build a second nest for the next year. Other bird species use Osprey nests by nesting in cavities or by usurping the nests before the Ospreys return after spring migration; others may nest on lower parts of the nest structures (Poole et al. 2002).

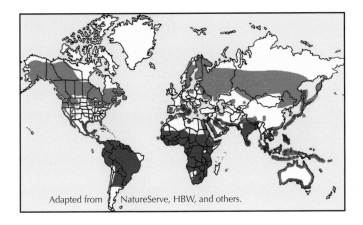

Adapted from NatureServe, HBW, and others.

NatureServe
Conservation Status

Global: .... Secure
Canada: .... Secure
US: .... Secure
Mexico: .... Secure

www.natureserve.org

Hook-billed Kite

Scientific: *Chondrohierax uncinatus*
Français: Milan bec-en-croc
México: Gavilán Pico de Gancho
Order: Accipitriformes
Family: Accipitridae

Human-made kites were named after this loosely-related group of raptors because of the similar flight patterns: sudden diving, rising, and twisting. The Hook-billed Kite can be difficult to find because it spends much of its time perched inside canopy foliage, where it moves little while foraging and laughing at birders below who see neither the bird nor the pile of snail shells on the ground. Once found, however, it can usually be scrutinized carefully because it is not wary. This extra identification time is important because of this species' highly variable coloration and bill size. Structural characteristics such as the unique small patch of yellow-orange skin in front of the eye, strongly hooked bill, and pinched-in rear wing-bases (seen in flight), are very helpful. Note the pale eye. This species occasionally soars, but usually not very high nor for very long.

The Hook-billed Kite feeds almost exclusively on tree snails. Its remarkable regional variation in bill size seems to have resulted from differences in snail size between regions. Regardless of region, bill size, or snail size, however, Hook-billed Kites apparently always use the same technique to extract snails from their shells. It was in this species that the technique was described for the first time—starting at the opening, the bird drives its hooked upper mandible

Gifi Beaton

Gavin Bieber

through successive whorls toward the apex of the shell's spiral. It then extracts the snail's soft body and eats it (Smith and Temple 1982).

The Hook-billed Kite colonized southern Texas from Mexico in recent decades and its population north of Mexico was thought to consist of 10–20 breeding pairs by the mid-1980s (Peterson 1960; Ridgely and Greenfield 2001; Bierregaard and Kirwan 2013).

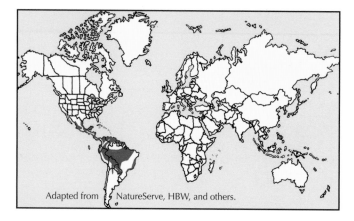

Adapted from NatureServe, HBW, and others.

NatureServe
Conservation Status

Global: ...... Apparently Secure
Canada: .... Not Applicable
US: .... Imperiled
Mexico: ..... Apparently Secure

www.natureserve.org

ROBERT ALVO

Swallow-tailed Kite

Scientific: *Elanoides forficatus*
Français: Milan à queue fourchue
México: Milano Tijereta
Order: Accipitriformes
Family: Accipitridae

The Swallow-tailed Kite is unmistakable with its pointed wings, very long, deeply forked tail, and striking black-and-white pattern. Its habitat requirements are mainly structural and consist of a group of tall, accessible trees for nesting, and nearby open areas that have enough small prey. Many habitats may satisfy these simple requirements. In Florida, although some other raptor species may use similar nesting habitat to that of the Swallow-tailed Kite, the latter usually forages in more open country than do the Red-shouldered Hawk and Short-tailed Hawk, but not in the extensive, treeless marshes inhabited by the Snail Kite. The Swallow-tailed Kite tends to use slightly drier, more continuously forested areas than does the Mississippi Kite. White-tailed Kites are found in areas that are more open with fewer large trees.

Insects are the mainstay of the adult Swallow-tailed Kite's diet throughout the year, but nestlings and recently fledged young are also fed small vertebrates. Parents often return to their nest with a whole wasp nest, dump out the larvae for their young, then sometimes recycle the papery material into their own nest structure. Fire ants are sometimes taken in large numbers without apparent damage to the kites, which have very thick and spongy stomach linings. The Swallow-tailed Kite captures all its prey with its feet while in flight by picking it off vegetation, reaching into or under foliage, or grabbing flying insects. It then flies with the food in its feet and transfers it to its mouth just before landing. This method often causes yellowish or greenish-brown staining of its lower white abdomen feathers by the prey.

David Laliberte

The US population of the Swallow-tailed Kite used to breed throughout Florida and the coastal region of the southeastern US, and in the Mississippi Valley north to Minnesota, but it declined dramatically from 1880 to 1940. Neither the population size nor the breeding distribution has ever rebounded. The two main causes of the decline were shooting and the reduction of appropriate forests, largely for agricultural use. The remaining 1000 or so pairs are now restricted to a much smaller range (Meyer 1995).

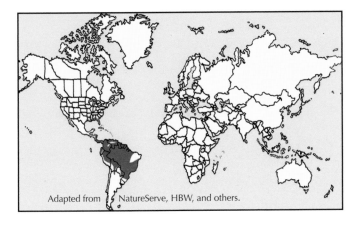

Adapted from NatureServe, HBW, and others.

NatureServe
Conservation Status

Global: .... Secure

Canada: .... Not Applicable

US: .... Vulnerable

Mexico: .... Secure

www.natureserve.org

White-tailed Kite

Scientific: *Elanus leucurus*
Français: Élanion à queue blanche
México: Milano Cola Blanca
Order: Accipitriformes
Family: Accipitridae

The White-tailed Kite, a small-mammal specialist feeder, lives in the gray zone. Whereas its US population was almost extirpated in the early 1900s, the range there has been expanding and is now wider than ever recorded. It is still not clear whether the New World form is a distinct species from two Old World forms, and biologists don't know whether it is a migrant, a nomad, or both. In addition, nests are sometimes located much closer to each other than is usually the case in raptors, leading to the debate of whether this species defends small breeding territories or whether it should be considered semi-territorial instead. Other evidence for the semi-territorial hypothesis includes the proximity of conspecifics during foraging, and the low aggressiveness shown toward them at any time.

Legs, feet, and talons are important features of this raptor in both actions and displays. "The leg-dangling habit of the kites is one of their most conspicuous oddities. On the nesting territory the protesting birds flew here and there nearly constantly, uttering their cries, beating the air slowly with short strokes, the wings held up at a sharp angle above the back, and the legs dangling from a point about the center of the body" (Bent 1937, 60). The White-tailed Kite dangles its legs in defense and in territorial behavior (to warn of its dangerous talons), while hovering (to help it hold its position), when dropping feet first onto prey, as well as during courtship display. Young birds and captured birds use a threat display in which they face the source of danger and lie back with their wings outstretched, talons spread, and mouth wide open. Males use their talons during copulation to stay perched on the female's back (Dunk 1995).

Steve Zamek

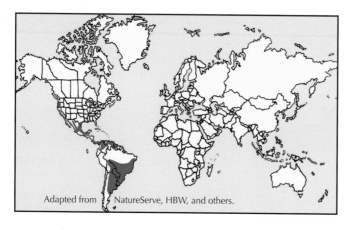

Adapted from NatureServe, HBW, and others.

NatureServe
Conservation Status

Global: .... Secure

Canada: .... Not Applicable

US: .... Apparently Secure

Mexico: .... Secure

www.natureserve.org

Snail Kite

Scientific: *Rostrhamus sociabilis*
Français: Milan des marais
México: Gavilán Caracolero
Order: Accipitriformes
Family: Accipitridae

A dietary specialist, the Snail Kite feeds itself and its young almost exclusively on freshwater apple snails of the genus *Pomacea*. In the US it breeds only in Florida, where it relies heavily on *Pomacea paludosa*. It searches for these snails in open parts of large marshes during flight or from a perch. Upon seeing one, the kite hovers just above the water surface and extends its feet down onto the snail. An accessible one may be moving, grazing, or resting on the bottom or on vegetation at or below the water surface down to a maximum depth of reach (for the kite) of 16 cm. The Snail Kite depends on water temperatures high enough for the snails to be close to the surface. Snail Kites and Limpkins both depend on *Pomacea paludosa* in Florida, but they partition the resource: the kites feed by sight mostly in open-water areas, whereas Limpkins feel for their food in more shallow areas that have dense emergent vegetation, dense floating vegetation, or open water.

The original threat to Snail Kites in Florida was, and continues to be, habitat loss. About one-quarter of peninsular Florida used to be flooded much of each year, providing ideal conditions for *Pomacea paludosa* and Snail Kites. Much of this area was drained from 1881 to the 1970s, and this caused a reduction in the kite's Florida range. The other major threat, a physical obstacle to feeding, is the formation of vegetation mats often consisting of invasive species whose growth is enhanced by fertilizers entering wetlands from surrounding farms and suburban developments.

The Snail Kite has an unusual mating system. During years with favorable foraging conditions, one parent deserts the other about one week before their young fledge. While the deserted mate continues to forage for the young for another three–five weeks, the deserter can breed again with another individual. Mate desertion does not occur in drought years (Sykes et al. 1995).

David Laliberte

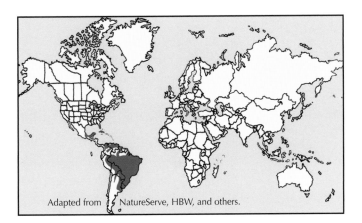

Adapted from NatureServe, HBW, and others.

NatureServe
Conservation Status

Global:...... Apparently Secure
Canada:.......... Not Applicable
US:...........Critically Imperiled
Mexico:..... Apparently Secure

www.natureserve.org

Mississippi Kite

Scientific: *Ictinia mississippiensis*
Français: Milan du Mississippi
México: Milano de Mississippi
Order: Accipitriformes
Family: Accipitridae

In the eastern part of their US range, Mississippi Kites nest primarily in old-growth forest, where they are not very colonial or abundant. In the Great Plains, however, they have often been plentiful in areas with many mature shelterbelts (lines of trees acting as windbreaks and erosion-controlling barriers). During the 1960s they began nesting in towns of all sizes in parts of Kansas, Oklahoma, Texas, and New Mexico, where they can be locally abundant and where they usually nest colonially. Urban breeding sites may have four–five nests per three-block area while other nearby, large areas of trees are unoccupied. Roosts of 10 or more birds are often situated near these groups of nests. Urban birds may land to drink water from ponds and puddles. Preferred urban breeding habitats are parks, residential areas, and golf courses where, in an effort to protect their nests, Mississippi Kites sometimes dive at humans who come too close. Diving individuals are a minority, and even fewer of those actually hit people, but the presence of a diving raptor is frightening even to the seasoned field biologist and poses a challenge to urban managers trying to appease both the average citizen and the conservationist (or animal rights activist).

A non-lethal but time-intensive solution is to remove the nest and move the nestlings to a foster pair. Alternatively, people can be advised simply to avoid active nests, wear a tall-crowned hat near nests, or wave something in front of a diving kite and keep moving.

Dennis Paulson

In contrast to protective breeding adults, Mississippi Kite nestlings are rarely aggressive. They may at times peck at each other mildly, but do not fight over food. Rather, they "work" by arranging nesting material and preening each other. Their low degree and frequency of aggression, as well as their work behavior, are uncommon in nestlings of other raptor species (Parker 1999).

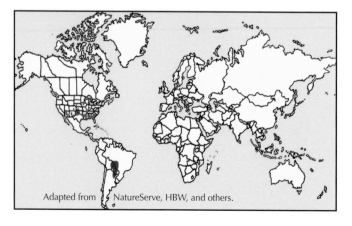

Adapted from NatureServe, HBW, and others.

NatureServe
Conservation Status

Global: .... Secure

Canada: .... Not Applicable

US: .... Secure

Mexico: .... Not Applicable

www.natureserve.org

Bald Eagle

Scientific: *Haliaeetus leucocephalus*
Français: Pygargue à tête blanche
México: Águila Cabeza Blanca
Order: Accipitriformes
Family: Accipitridae

The Bald Eagle is the national emblem of the US and a symbol of strength, freedom, and wilderness. It roams over large bodies of water such as sea coasts, rivers, lakes, and, more recently, reservoirs. Fish are its preferred food type, but it is an opportunistic forager that also eats mammals, birds, and reptiles. It scavenges prey items when available, pirates food from other birds and mammals, and also captures its own prey. When foraging, it favors shallow waters, taking advantage of situations in which it can find killed, moribund, or stunned prey, such as the outflows of hydroelectric power plants. In spring, Bald Eagles and other scavengers find fish kills at shallow lakes and ponds that had become oxygen-depleted in late winter under their ice cover. Other favorite situations are shallow, riverine gravel-bars full of dead salmon after spawning and, in winter, areas of waterfowl concentration with birds killed by hunters.

Coyotes (*Canis latrans*) invaded the island of Newfoundland in the 1980s, arriving on ice floes. They prey on Caribou (*Rangifer tarandus*), but leave its carcasses partly eaten. This has helped Bald Eagles, which relied on the Atlantic Cod (*Gadus morhua*) industry for fish offal until imposition of a moratorium on that fishery in 1992.

Juan Bahamon

Bald Eagle nests are among the largest bird nests in the world. Typically they are 1.5–1.8 m in diameter and 0.7–1.2 m tall. In order to fit in the selected tree, they must conform to its shape, and thus range from cylindrical to cone-shaped to flat. Nests are also built on cliffs. The largest nest reported was 2.9 m in diameter and 6.1 m tall. Bald Eagles have been seen playing with objects and passing sticks to each other in the air, perhaps sharpening their nest-building or foraging prowess (Buehler 2000).

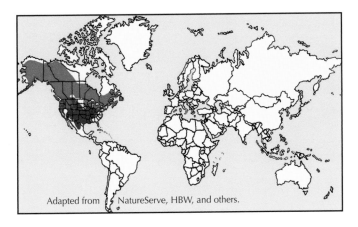

Adapted from NatureServe, HBW, and others.

NatureServe
Conservation Status

Global: .... Secure

Canada: .... Secure

US: .... Secure

Mexico: .... Critically Imperiled

www.natureserve.org

Northern Harrier

Scientific: *Circus cyaneus*
Français: Busard Saint-Martin
México: Gavilán Rastrero
Order: Accipitriformes
Family: Accipitridae

The Northern Harrier is different from most of our other raptors (both diurnal and nocturnal) in its high degree of sexual dimorphism in plumage and in its tendency to be polygynous. With its disk-shaped face, it probably relies more on hearing to catch prey than do other diurnal raptors. It is more closely associated with the ground, both nesting and roosting there, and it quarters low over the ground while hunting for small mammals and small birds. Also unlike most of our other raptors, migrating Northern Harriers fly in light rain and snow. They are less affected by weather patterns such as the timing of cold fronts because they fly lower and use ground-layer air currents more than do other raptors, and they use more flapping and less soaring and gliding (Allen et al. 1996).

On the other hand, from an ecological viewpoint, the Northern Harrier is very similar to the unrelated Short-eared Owl. Both are ground-nesters and feed in treeless habitats such as marshes and fields, where they forage in a similar pattern and take much the same prey. Where they co-occur, their territories may overlap, but they partition the foraging habitat by time of day. When they meet, the harrier may steal food from the owl (Smith et al. 2011).

Northern Harrier nests constructed in dry areas with little chance of flooding may be only a few centimeters above the ground, whereas those in areas with a high probability of flooding can be structures built up from the ground to 45 cm (Bent 1937).

Carol Blackard

Tom Whetten

Edgar T. Jones

Robert Alvo

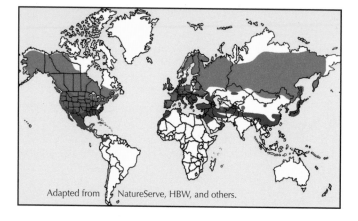
Adapted from NatureServe, HBW, and others.

NatureServe
Conservation Status

Global: .... Secure

Canada: .... Secure

US: .... Secure

Mexico: ..... Apparently Secure

www.natureserve.org

Sharp-shinned Hawk

Scientific: *Accipiter striatus*
Français: Épervier brun
México: Gavilán Pecho Canela
Order: Accipitriformes
Family: Accipitridae

The English word "accipiter" refers to birds of the genus *Accipiter*. These species make up only a small portion of the family Accipitridae and of course even a smaller portion of the order Accipitriformes. Our three accipiters are the Sharp-shinned Hawk, Cooper's Hawk, and Northern Goshawk. All three are slim with short, broad, rounded wings and long tails, which allow them to maneuver easily in forests and shrublands to ambush small birds and mammals after a short chase. Their flight of several flaps followed by gliding helps to distinguish them from other raptors.

The Sharp-shinned Hawk is "the terror of all small birds and the audacious murderer of young chickens" having "been well-called a bushwacker from its habit of beating stealthily about the shrubbery to the fatal surprise of many a little songster" (Bent 1937, 95). Being very secretive, it breeds in large forest stands where its big eyes help it see in low light conditions.

Julie Dufour

Its long middle toes help it reach through vegetation to seize prey and retain it with its deeply curved, needle-like talons while both predator and prey are moving quickly.

The Sharp-shinned Hawk is most easily observed during migration. It comprised 12–33% of all raptors seen at three major North American autumn "hawk watches". This relative abundance, its tendency to migrate along traditional corridors, and its habit of hunting on migration all contribute to making it one of the migrating North American raptors that is most frequently trapped for leg banding. Resulting data show that males average 57% the weight of females, the lowest such fraction in all North American raptors.

Winter is also a good time to observe Sharp-shinned Hawks due to their habit of feeding on birds in human-modified areas, especially at bird feeders, where they are responsible for about 35% of observed predation (Dunn and Tessaglia 1994; Bildstein and Meyer 2000).

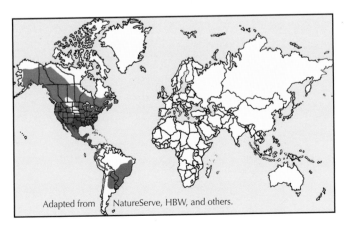

Adapted from NatureServe, HBW, and others.

NatureServe
Conservation Status

Global: .... Secure

Canada: .... Secure

US: .... Secure

Mexico: .... Secure

www.natureserve.org

Cooper's Hawk

Scientific: *Accipiter cooperii*
Français: Épervier de Cooper
México: Gavilán de Cooper
Order: Accipitriformes
Family: Accipitridae

"If the Sharp-shinned Hawk is a blood-thirsty villain, this larger edition of feathered ferocity is a worse villain, for its greater size and strength enable it to do more damage.... It is essentially *the* chicken hawk so cordially hated by poultry farmers, and is the principal cause of the widespread antipathy toward hawks in general" (Bent 1937, 112). How times have changed! While Cooper's Hawks still take some poultry, hawk-shooting hills like Hawk Mountain, Pennsylvania, have morphed into "hawk watch" sites for migration monitoring.

Cooper's Hawk numbers started declining in the early 1900s due to shooting. This was followed by a major decline after World War II that was detected using migration counts. Its main cause was the use of pesticides, especially DDT, which moved up the food chain and inhibited reproduction in numerous raptor species by causing eggshell thinning followed by reproduction failure. After the widespread use of DDT was banned in the US in 1972, the Cooper's Hawk population rebounded. By the 1990s the species had colonized urban, suburban, and recreational areas, as well as plantations, all of which may have plenty of medium-sized birds (the hawk's main prey). These unnatural habitats mimic the species' natural foraging habitat in having human-made obstacles used for ambushing prey. Doves

Larry Master

apparently make up 57% of prey brought to nestlings at urban nests, but only 4% at rural nests. A disadvantage to urban Cooper's Hawks may be a high death rate of nestlings from trichomoniasis obtained from doves. Another consequence of urban life is collisions with cars and other moving objects. The main unknown in this "villain's" success story is the effect its relatively large numbers may be having on prey species (e.g., American Kestrel) and competitors, especially those in decline (Curtis et al. 2006).

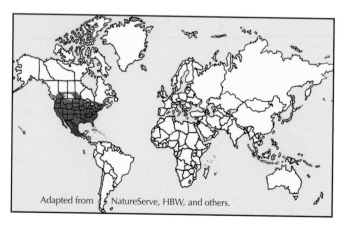

Adapted from NatureServe, HBW, and others.

NatureServe
Conservation Status

Global: .... Secure

Canada: .... Secure

US: .... Secure

Mexico: .... Secure

www.natureserve.org

ROBERT ALVO

Northern Goshawk

Scientific: *Accipiter gentilis*
Français: Autour des palombes
México: Gavilán Azor
Order: Accipitriformes
Family: Accipitridae

This species is notorious for its extreme aggressiveness toward intruders near its nest, especially when it has early-stage nestlings. Attacks on one person are usually more violent than those on two or more people, and the bird may even draw blood. It is less aggressive toward human intruders in Europe, possibly because it has had a longer period of persecution in which to learn to fear us.

It is "the largest, handsomest, and most dreaded of the *Accipiter* tribe" (Bent 1937, 125), generally taking larger prey (mammals and birds typically weighing 250–450 g) than do the smaller Sharp-shinned and Cooper's Hawks. Surprise is of the essence for accipiter foraging success. The Northern Goshawk's prey occurs mostly in the ground- and shrub-zones. If the hawk goes undetected it may descend quietly without beating its wings, in an accelerating glide onto its prey. However, if it sees that it has been detected it flaps hard, often crashing recklessly through shrubs and other vegetation, even entering water in the chase if necessary. But if the hawk is undetected, it starts its final glide when 8–9 m from its target. At 1–3 m it lowers its feet with toes partially flexed, and then extends the legs in front. At 0.7–0.9 m it fully extends its toes. This hawk kills its prey by driving its talons into it and kneading it. Its quarry may then be cached in a nearby tree, especially if early-stage nestlings are present, for the parent must feed them frequently. Processing of Northern Goshawk prey into a pellet takes about 21 hours (Squires and Reynolds 1997).

Edgar T. Jones

Larry Master

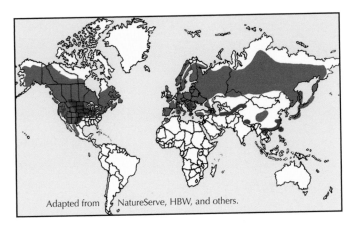

Adapted from NatureServe, HBW, and others.

NatureServe
Conservation Status

Global: .... Secure

Canada: .... Secure

US: .... Apparently Secure

Mexico: ..... Apparently Secure

www.natureserve.org

Common Black Hawk

Scientific: *Buteogallus anthracinus*
Français: Buse noire
México: Aguililla Negra Menor
Order: Accipitriformes
Family: Accipitridae

South of the US, the Common Black Hawk nests in a number of habitats, typically aquatic ones. In the southwestern US, however, it nests exclusively along streams, preferring mature forests along perennial streams in otherwise treeless areas. Typical streams that it uses are less than 30 cm deep with riffles and perches such as branches, boulders, ledges, and sandbars. It will also use streams that are partly dry as long as they have some pools or seepage areas. This bird is an opportunist, taking small to medium-sized vertebrates and invertebrates. It prefers fish, crayfish, and other aquatic animals. South of the US, crustaceans (e.g., crayfish) are sometimes the only prey that it eats.

The Common Black Hawk usually hunts from a perch, gliding down onto its prey, or it may move to a perch close to its quarry and then drop to seize it with its talons. It also hunts by walking along the stream, hopping from rock to rock aided by some flapping, or by flying below the canopy from limb to limb or from rock to rock. It may fly over the stream and snatch a fish from the surface. Another foraging method involves wading in a pool and waving its wing tips through the water to move prey to the shallows. Competitors like herons or kingfishers may be chased away from its stream.

Dennis Paulson

Only 220–250 breeding pairs are estimated to exist in the US, and the population seems to be occupying all the available riparian habitat there. The greatest threats to the population are clearing or alteration of its habitat, water diversion from streams, diking and damming of streams to control flooding, lowering of the water table by overusing well-water, and livestock grazing at unfenced streams. Also, the exotic invasive plant Salt Cedar (*Tamarix chinensis*) outcompetes native nest trees and is too short for nesting. The top management priority for this hawk is to conserve, rejuvenate, or rehabilitate its habitat as appropriate. Streams with their surrounding forests are ecosystems deserving protection, and this hawk could perhaps be used as their symbol in its US range (Schnell 1994).

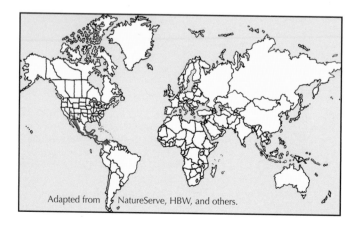

Adapted from NatureServe, HBW, and others.

NatureServe
Conservation Status

Global: ...... Apparently Secure
Canada: .... Not Applicable
US: .... Vulnerable
Mexico: ..... Apparently Secure

www.natureserve.org

ROBERT ALVO

Harris's Hawk

Scientific: ***Parabuteo unicinctus***
Français: Buse de Harris
México: Aguililla Rojinegra
Order: Accipitriformes
Family: Accipitridae

Long before Europeans arrived in North America, First Nations people were aware of the Harris's Hawk's unique social nature, the details of which were studied extensively by biologists starting in the 1970s. Social units have a hierarchy and are made up of a breeding pair plus one to five helpers that may or may not be relatives. Helpers assist with incubation and in brooding, shading, feeding, and playing with the nestlings. Helping may co-occur with monogamy, polyandry, and/or polygyny. Social units generally avoid each other rather than act aggressively as do other raptors, and nests of different social units are usually at least 500 m apart.

The Harris's Hawk is the only North American raptor that regularly hunts in groups, its highly sophisticated cooperative tactics marking the pinnacle of those known in birds. The birds first "meet", sometimes with the entire unit perched on one branch. Then they take advantage of their numbers: one bird flushes the prey (e.g., a rabbit) while others catch it; one bird takes over the chase when another has tired; or several birds attack the prey from different directions to cut off escape routes. Once they've caught and killed the prey, they circulate it until the whole unit has fed or taken pieces up to perches. If they have a nest, they then feed the nestlings. Group hunting is known in North America but not in Latin America.

Harris's Hawks have been colonizing cities in Arizona and Chile since the 1980s. Their social nature sometimes leads to electrocution of birds when they touch differentially energized conductors on electric power poles during food transfers or pre-hunting meetings, or when its members are perched simply to demonstrate their rank in the social unit according to each bird's height above the ground (Dwyer and Bednarz 2011).

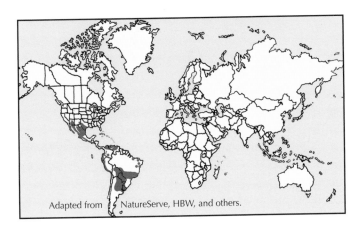

Juan Bahamon

Adapted from NatureServe, HBW, and others.

Red-shouldered Hawk

Scientific: *Buteo lineatus*
Français: Buse à épaulettes
México: Aguililla Pecho Rojo
Order: Accipitriformes
Family: Accipitridae

"Sportsmen, farmers, and poultry- and game-breeders are all sworn enemies of all hawks and will not be convinced that there is any good hawk but a dead hawk. The bounty system is far too prevalent and leads to the killing of far too many old and young hawks in or near their nests, which the farmers hunt up and watch until the young hatch; the old birds are then more easily shot and the heads of the young secured. I believe we have saved the lives of many a family of hawks by taking the eggs in April. They lay a second set in May and stand a better chance of raising a brood when the leaves are out, for the nests are harder to find and the farmers have ceased to look for them" (Bent 1937, 195). So it seems that Bent and perhaps some colleagues ("we") removed eggs from Red-shouldered Hawk nests to delay the breeding process and enhance the birds' chances of success while state governments were paying people good money to shoot any raptor. The entire group of species was considered "bad" while other birds (e.g., many songbirds) were labelled "good". Myths abounded. For example, Bald Eagles were said to carry human babies to their nests, whereas smaller species such as Red-shouldered Hawks supposedly consumed large numbers of farm animals such as poultry. Bounties are too simplistic—they often don't work, are very expensive, and/or often have unintended bad consequences. The poor reputation of raptors, on which bounties were based, was so widespread that the US Migratory Bird Treaty Act of 1918 didn't protect them until it was amended in 1972. Falconers have been raptor conservation's long-time proponents (Dykstra et al. 2008).

Kelly A. Boadway

Larry Master

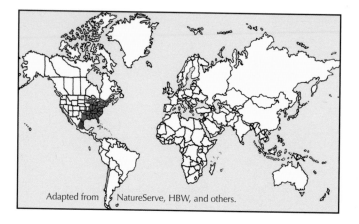
Adapted from NatureServe, HBW, and others.

NatureServe
Conservation Status

Global:........................ Secure

Canada:..... Apparently Secure

US:.............................. Secure

Mexico:....................... Secure

www.natureserve.org

Broad-winged Hawk

Scientific: *Buteo platypterus*
Français: Petite Buse
México: Aguililla Alas Anchas
Order: Accipitriformes
Family: Accipitridae

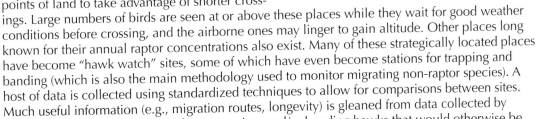

Raptors migrate during the day, probably because that is when the best wind conditions for fast flight mostly occur. Other birds, however, migrate mostly at night, probably to avoid hungry raptors. Birders can thus watch, identify, and count raptors during migration, which they cannot do with most other birds.

Migrating raptors avoid crossing open water because the thermals they use for soaring are not as strong there as the ones they find over land, so they often concentrate at peninsulas and other points of land to take advantage of shorter cross-ings. Large numbers of birds are seen at or above these places while they wait for good weather conditions before crossing, and the airborne ones may linger to gain altitude. Other places long known for their annual raptor concentrations also exist. Many of these strategically located places have become "hawk watch" sites, some of which have even become stations for trapping and banding (which is also the main methodology used to monitor migrating non-raptor species). A host of data is collected using standardized techniques to allow for comparisons between sites. Much useful information (e.g., migration routes, longevity) is gleaned from data collected by

Giff Beaton

watching, trapping, and/or banding hawks that would otherwise be difficult to obtain for this generally secretive group of birds. Some of this information is difficult to obtain for non-raptors, such as the number of birds of each species migrating per night over a site.

The Broad-winged Hawk migrates in "kettles" ranging from a few birds to thousands. Like many other raptors, this species typically migrates by soaring up in a thermal, then gliding down to the next one. It generally resorts to flapping flight only during rain and early or late in the day. Recent migration counts in eastern Mexico establish a minimum autumn estimate of the Broad-winged Hawk's global population at 1.7 million birds, not including its five nonmigratory Caribbean subspecies (Goodrich et al. 2014).

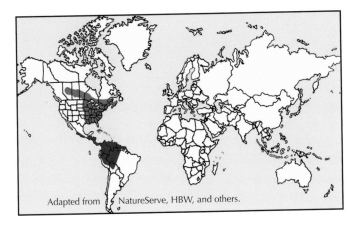

Adapted from NatureServe, HBW, and others.

NatureServe
Conservation Status

Global: .... Secure

Canada: .... Secure

US: .... Secure

Mexico: .... Not Applicable

www.natureserve.org

Gray Hawk

Scientific: *Buteo plagiatus*
Français: Buse grise
México: Aguililla Gris
Order: Accipitriformes
Family: Accipitridae

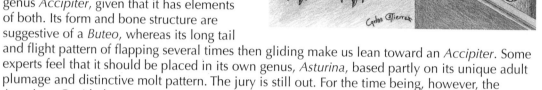

What is it? Well, it's a diurnal raptor of some sort, and probably a hawk. Indeed, most studies of the Gray Hawk have examined its taxonomic status, specifically whether it belongs in the genus *Buteo* or the genus *Accipiter*, given that it has elements of both. Its form and bone structure are suggestive of a *Buteo*, whereas its long tail and flight pattern of flapping several times then gliding make us lean toward an *Accipiter*. Some experts feel that it should be placed in its own genus, *Asturina*, based partly on its unique adult plumage and distinctive molt pattern. The jury is still out. For the time being, however, the American Ornithologists' Union considers it a *Buteo*.

The small US Gray Hawk population probably consists of fewer than 100 breeding pairs limited to Arizona, New Mexico, and Texas. Like the Common Black Hawk, this is a riparian-nesting species, although rather than foraging in and around streams, it does so in nearby mesquite (*Prosopis spp.*) woodlands. Since the late 1800s, when logging began having considerable habitat impacts in these states, Gray Hawk numbers and distribution have shifted according to the loss of riparian vegetation in some areas, then the creation of new habitats elsewhere. Recent work has shown that Gray Hawks can reach high densities in good habitat. This suggests—along with the fact that populations in Arizona and Texas seem to be stable or increasing, plus recent observations of Gray Hawks nesting at smaller tributaries—that this species should continue to breed in the US for some time, even though groundwater depletion poses a threat (Bibles et al. 2002).

Carol Blackard

Christian Artuso

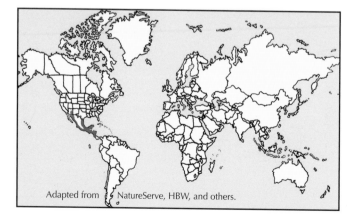

Adapted from NatureServe, HBW, and others.

NatureServe
Conservation Status

Global: .... Not Yet Ranked

Canada: .... Not Applicable

US: .... Vulnerable

Mexico: .... Secure

www.natureserve.org

Short-tailed Hawk

Scientific: ***Buteo brachyurus***
Français: Buse à queue courte
México: Aguililla Cola Corta
Order: Accipitriformes
Family: Accipitridae

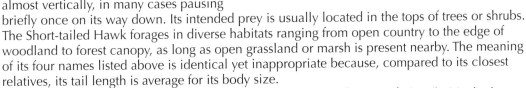

The main distinguishing characteristic of the Short-tailed Hawk is its aerial hunting behavior of kiting (hanging motionless) on spread wings while facing into the wind and scanning the ground directly below for small to medium-sized birds. When it dives at prey, it drops almost vertically, in many cases pausing briefly once on its way down. Its intended prey is usually located in the tops of trees or shrubs. The Short-tailed Hawk forages in diverse habitats ranging from open country to the edge of woodland to forest canopy, as long as open grassland or marsh is present nearby. The meaning of its four names listed above is identical yet inappropriate because, compared to its closest relatives, its tail length is average for its body size.

The isolated US population is separated from the nearest other population (in Mexico's Yucatan Peninsula) by the Gulf of Mexico, some 800 km by direct flight. No individuals are known to have traveled from one region to the other, and this segregation is thought to be 20,000 years old. The US population is probably endemic to Florida, where it spreads out over most of the peninsula to breed, but then moves south in a well-defined migration to winter in the southern peninsula and Florida Keys. Population estimates for Florida are 150–200 breeding pairs or at most 500 individuals (including nonbreeders). The species seems to have always been rare in the US and apparently no major changes in abundance or distribution have occurred over the past 60 years or so. The main reasons for its Imperiled US status are its small population size, its isolation, and ongoing degradation of its habitat.

The Short-tailed Hawk has light and dark color morphs that interbreed. Both sexes occur as both color morphs. Dark birds outnumber light ones in the US, whereas the opposite is true in Central and South America (Miller and Meyer 2002; Maehr and Kale 2005).

David Laliberte

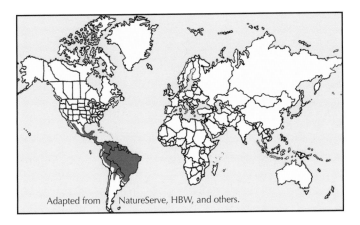

Adapted from NatureServe, HBW, and others.

NatureServe
Conservation Status

Global: ...... Apparently Secure

Canada: .... Not Applicable

US: .... Imperiled

Mexico: ..... Apparently Secure

www.natureserve.org

Swainson's Hawk

Scientific: ***Buteo swainsoni***
Français: Buse de Swainson
México: Aguililla de Swainson
Order: Accipitriformes
Family: Accipitridae

Outside of the breeding season, the Swainson's Hawk is different from most of our other hawks in that it forages almost exclusively on insects, especially grasshoppers. It is also the most gregarious of our breeding raptors. On the wintering grounds, for example, foraging assemblages and nocturnal roosts may include thousands of individuals. Even during the breeding season, nonbreeding birds in central California may form foraging flocks of more than 100 birds and use communal, nocturnal roosts. In late August and early September, flocks of several hundred Swainson's Hawks, both immatures and adults, gorge on grasshoppers to lay down fat for the 10,000-km migration south to the pampas of South America, especially those in Argentina. The birds migrate overland during the day in flocks of up to 10,000 birds. Most of the population passes through Central America in a relatively short (two-week) period in both autumn and spring, creating a spectacular event that can be seen during the day.

Swainson's Hawk migration was first studied using satellite telemetry in 1995, and led to the discovery that an estimated 20,000 of these hawks perished in Argentina in 1995 and 1996. In contrast to the eggshell thinning that was found in various raptor species in North America during the DDT era, the problem on the Swainson's Hawk's wintering grounds was acute toxicity from the spraying of the organophosphate insecticides monocrotophos and dimethoate to control grasshopper outbreaks in sunflower and alfalfa fields. Some birds died minutes after being sprayed while foraging, whereas others died several days after eating sprayed grasshoppers. Monocrotophos was banned in Argentina in 1999, but was replaced by other highly toxic pesticides (Bechard et al. 2010).

Christian Artuso

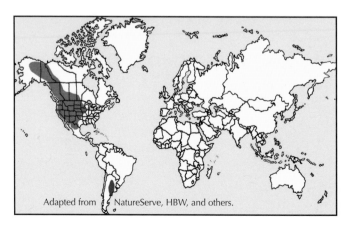
Adapted from NatureServe, HBW, and others.

NatureServe
Conservation Status

Global: .... Secure

Canada: ..... Apparently Secure

US: .... Secure

Mexico: .... Secure

www.natureserve.org

White-tailed Hawk

Scientific: *Buteo albicaudatus*
Français: Buse à queue blanche
México: Aguililla Cola Blanca
Order: Accipitriformes
Family: Accipitridae

As per this species' four names listed above, adults have a white tail, though it has a black band near the tip. This hawk, thought to number only a few hundred pairs in the US, does best in arid regions that are open or that have scattered trees or shrubs.

In one fascinating incident, two adult birds attacking a jackrabbit were surprised in a field by a passing tractor. The birds sat on nearby fence posts for two hours while the tractor cut and flattened the tall weeds. Immediately after the tractor departed, the birds flew up from their posts, and easily caught and ate the no-longer-concealed rabbit.

White-tailed Hawks attend fires in larger numbers than do other raptors, often with 3–20 individuals, but rarely with 60 or more. "As a management measure, a 150-acre (59 ha) tract was burned on January 18, 1939.... Some 36 raptors of six species arrived. This number included 16 adult and four immature White-tails. From our knowledge of White-tail distribution, it was our opinion that the fire attracted all, or nearly all, the adults present within a radius of 10 miles (16 km). The White-tails coursed back and forth parallel to the fire line and, at times, dived through the smoke for cotton rats, pocket mice, and grasshoppers" (Stevenson and Meitzen 1946, 204).

White-tailed Hawks are considered nonmigrants, yet unconvincing reports of migration have been made by a number of researchers. Perhaps this species at times joins true migrants temporarily in search of food (Farquhar 2009).

Lee Zieger

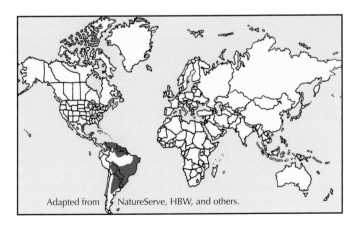

Adapted from NatureServe, HBW, and others.

NatureServe
Conservation Status

Global: ...... Apparently Secure

Canada: .... Not Applicable

US: .... Vulnerable

Mexico: ..... Apparently Secure

www.natureserve.org

Zone-tailed Hawk

Scientific: *Buteo albonotatus*
Français: Buse à queue barrée
México: Aguililla Aura
Order: Accipitriformes
Family: Accipitridae

Current thinking is that this hawk is an impostor. No one has tested this hypothesis yet, but it is backed to some degree by observational information. The Zone-tailed Hawk's global range easily falls within that of the very common and widespread Turkey Vulture, and a good way of finding a Zone-tailed Hawk in its small US range is to carefully check all the soaring Turkey Vultures above you. Just like a Turkey Vulture, the Zone-tailed Hawk holds its wings in a dihedral (looking like a "V" from head-on) and teeters from side to side in flight, appearing unstable. The coloring and shape are also generally similar to the Turkey Vulture's. However, the hawk is smaller, has a broader, dark-feathered head (instead of a red, unfeathered one), has a narrow dark trailing edge of its underwings, has more bulging secondaries, and often shows several light bands on the tail. Prey species often become

Dennis Paulson

used to seeing many soaring Turkey Vultures, and are not afraid of them. When an impostor sees prey, it keeps circling until it goes out of view behind an obstacle such as a rock outcrop or a shrub, then emerges from cover and dives after its prey. Zone-tailed Hawks in the US often nest within 0.8 km of Turkey Vulture roosts, and fly and forage during the same daytime hours as the vultures. The hawks even roost among the vultures.

Another hypothesis requiring testing is that a commensal relationship (where one organism benefits without affecting the other) exists between two species of hummingbird (the Black-chinned and Broad-tailed), on the one hand, and the Zone-tailed Hawk on the other. These hummingbirds regularly visit this hawk's nests that contain young and feed on insects attracted by uneaten prey (Johnson et al. 2000).

Johnson et al. (2000) recount some excellent old stories of researchers risking their lives to collect this species' eggs.

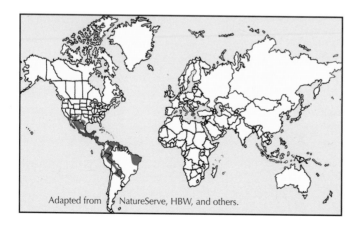

Adapted from NatureServe, HBW, and others.

NatureServe
Conservation Status

Global: ...... Apparently Secure
Canada: .... Not Applicable
US: .... Apparently Secure
Mexico: ..... Apparently Secure

www.natureserve.org

Red-tailed Hawk

Scientific: ***Buteo jamaicensis***
Français: Buse à queue rousse
México: Aguililla Cola Roja
Order: Accipitriformes
Family: Accipitridae

"The curious habit of the old birds in gathering a green leafy bough and placing it in the nest, characteristic of the Swainson's Hawk also, is very marked in the Red-tailed Hawk, a fresh bough being gathered at least once daily during the time when the young are small. There has been some doubt hitherto as to the cause of this habit, but by observing the nestlings I am led to believe that the bough acts as a sunshade, as the young have been seen to repeatedly pull the bough over themselves and crouch beneath it. Doubtless it also acts as a shield and hides the young from their enemies" (Criddle 1917, 75).

The Red-tailed Hawk is a bird of open habitats dotted with trees or groups of trees. Human-made structures such as power line poles can substitute for trees, and are used by this sit-and-wait predator as nesting sites and perches from which to hunt. This habitat preference has served it well, for during the 20th century it expanded its range and replaced the Red-shouldered Hawk in much of eastern North America, and Swainson's and Ferruginous Hawks in parts of the Great Plains. It was the fragmentation of large expanses of forest by logging in the east and the fragmentation of prairies by fire-suppression in the west that aided this species while working to the detriment of others. It is no surprise that the Red-tailed Hawk population is Secure in the US, in Canada, and globally, for it is the commonest *Buteo* in most of North America.

In the tropics, the Red-tailed Hawk nests in high densities in closed-canopy rain forests and cloud forests, such as in Puerto Rico. There it dives on prey from the air far above the canopy (Preston and Beane 2009).

Raptor identification can provide hours of distraction while driving between birding sites because these birds are large and often perch on trees and poles next to the road. The common Red-tailed Hawk can easily be identified from behind, and thus eliminated as a possible less common species, by the white "V" on its wings.

Juan Bahamon

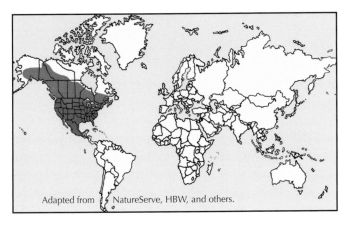

Adapted from NatureServe, HBW, and others.

NatureServe
Conservation Status

Global: .... Secure
Canada: .... Secure
US: .... Secure
Mexico: .... Secure

www.natureserve.org

Ferruginous Hawk

Scientific: *Buteo regalis*
Français: Buse rouilleuse
México: Aguililla Real
Order: Accipitriformes
Family: Accipitridae

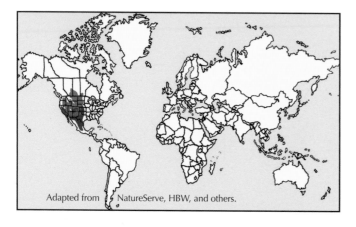

CAN YOU SIT ON THE EGGS EMILY? I'M A LITTLE TIED UP AT THE MOMENT...

DITTO HARRY...

The regal Ferruginous Hawk is true to its scientific name. Its French name (meaning "rusty") and its English name refer to its coloration. It is an open-country species of western North America's Great Plains, inhabiting grasslands and arid and semi-arid regions. Mammals such as rabbits, ground squirrels, and prairie dogs make up most of its prey.

The largest of our 10 *Buteos*, it is often compared to the larger Golden Eagle. "One who knows it in life cannot help being impressed with its close relationship to the Golden Eagle, which is not much more than a glorified *Buteo*" (Bent 1937, 284). Experienced falconers like working with it because it is larger and more powerful than other *Buteos*, and it yields a hunting experience almost as thrilling as that offered by the Golden Eagle with less risk of injury to the falconer by the hawk. More agile than the eagle, the Ferruginous Hawk takes large birds more easily.

At the global scale, this species is widespread and reasonably common in its habitat. On the other hand, it has declined in some regions and is sensitive to human disturbance. Furthermore, good habitat continues to be lost and its numbers are rather low, so there is some concern for this species as indicated by its status of Apparently Secure at the global, Canada, and US levels.

One of the main concerns, oddly enough for a bird species, is invasion by two exotic plants: Cheatgrass (*Bromus tectorum*) and Russian Thistle (*Salsola iberica*) (NatureServe 2014b). Both invasives grow so densely that they may hinder populations of mammalian prey, or make it difficult for Ferruginous Hawks to access their prey.

Juan Bahamon

Adapted from NatureServe, HBW, and others.

NatureServe
Conservation Status

Global: ...... Apparently Secure

Canada: ..... Apparently Secure

US: .... Apparently Secure

Mexico: .... Not Applicable

www.natureserve.org

Rough-legged Hawk

Scientific: ***Buteo lagopus***
Français: Buse pattue
México: Aguililla Ártica
Order: Accipitriformes
Family: Accipitridae

Carlos Gtierres

What a strange coincidence that Rough-legged Hawks have legs that resemble those of lagomorphs (rabbits and hares), one of their prey. Add to that reports of dead migrant Rough-legged Hawks that have been hit by cars on Utah highways while feeding on road-killed rabbits, and you might imagine a strange, macabre scene. After all, the name *lagopus* means "hare-footed".

The Rough-legged Hawk breeds over the huge taiga and tundra regions of the Northern Hemisphere. It seeks similar open habitats in temperate regions during migration and in winter. It prefers cliffs for nesting, but will sometimes use trees or human-made structures such as bridges, towers, and cairns. Cliff nests are usually built near the top.

I was very fortunate to find one of these nests in central Quebec at the southern extent of the species' breeding range. This region lies at the current northern limit of eastern North America's road network. When I visited in 1993 to scout the avifauna, it required a 1200-km drive north from Ottawa. The last stretch of 620 km had only one gas station, at kilometer 381. That paved road had been constructed in the 1970s to service seven huge reservoirs and associated

Dennis Paulson

hydroelectric power plants along the 900-km-long La Grande (LG) River. Another 700-km road (gravel) gave access to the power plants. In the instance of too much water in the LG2 reservoir, a spillway, "The Staircase of the Giants", was created by cutting out the hillside. It was 1 km long, 135 m wide, and shaped with 10 steps that were each 10 m high, down which any released water could fall and lose kinetic energy to avoid major landscape damage along the river's remaining 118 km. It was in the spillway's roughly 40-m-high cliffs that a Rough-legged Hawk nested. It let me know this by flying across the spillway to the opposite cliff to dive-bomb me. Having never seen a dark-phase Rough-legged Hawk, I spent some time poring unsuccessfully over my field guide, not realizing that the bird was the most common raptor species of the far north.

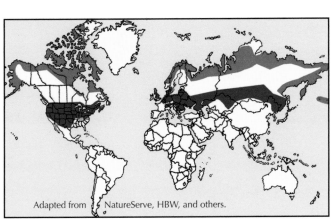

Adapted from NatureServe, HBW, and others.

NatureServe
Conservation Status

Global: .... Secure
Canada: .... Secure
US: .... Secure
Mexico: .... Not Applicable

www.natureserve.org

Golden Eagle

Scientific: *Aquila chrysaetos*
Français: Aigle royal
México: Águila Real
Order: Accipitriformes
Family: Accipitridae

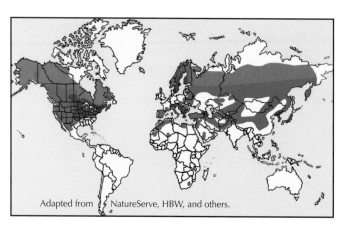

"This magnificent eagle has long been named the 'king of birds', and it well deserves the title. It is majestic in flight, regal in appearance, dignified in manner, and crowned with a shower of golden hackles about its royal head. When falconry flourished in Europe the Golden Eagle was flown only by kings" (Bent 1937, 293). The high-flying Golden Eagle was often shot from airplanes and helicopters in the mid-1900s by ranchers and gunners hired by ranchers, for these birds take domestic lambs in parts of the US west during lambing season (Kochert et al. 2002).

"The eagle was devouring the carcass of a blue hare when a fox sprang from the surrounding heather and seized the great bird by the wing. A well-contested struggle ensued in which the eagle made a desperate attempt to defend itself with its claws and succeeded in extricating itself from its enemy's grasp, but before it had time to escape, Reynard [the fox] seized it by the breast and seemed more determined than ever. The eagle made another attempt to overpower its antagonist by striking with its wings, but that would not compel the aggressor to quit its

Juan Bahamon

hold. At last the eagle succeeded in raising the fox from the ground, and for a few minutes Reynard was suspended by his own jaws between heaven and earth.... He soon found that the strong wings of the eagle were capable of raising him and that there was no way of escape unless the bird should alight somewhere. The eagle made a straight ascent and rose to a considerable height in the air. After struggling for a time Reynard was obliged to quit his grasp, and descended much quicker than he had gone up. He was dashed to the earth where he lay struggling in the agonies of death. The eagle made his escape, but appeared weak from exhaustion and loss of blood" (Gordon 1915). Such observations are rare gems.

Adapted from NatureServe, HBW, and others.

NatureServe
Conservation Status

Global: .... Secure
Canada: ..... Apparently Secure
US: .... Secure
Mexico: .... Critically Imperiled

www.natureserve.org

Yellow Rail

Scientific: *Coturnicops noveboracensis*
Français: Râle jaune
México: Polluela Amarilla
Order: Gruiformes
Family: Rallidae

The rare Yellow Rail has provided some of my favorite ornithological experiences. In the 1990s, Canadian Wildlife Service biologist Michel Robert hired me to co-write a national status report with him: it was a perfect occasion for learning about the species. Later, a calling male was heard near Quebec City, and Michel needed gut samples for a diet study. Armed with a bright lamp, a net, and two stones (for imitating its typewriter-like *tic-tic, tic-tic-tic* call), we caught the bird. By injecting the harmless fluid "tartar emetic" down its throat, we obtained a gut sample for food analysis. I couldn't believe the bird's tiny size. On another occasion, we used a hunting dog trained with Yellow Rail scent to look for nests. Doing this during the day, I appreciated the very shallow nature of the water in the rail habitat and its susceptibility to being drained by humans.

Years later I visited Manitoba's Douglas Marsh, a well-known breeding site. After five weeks of birding alone, I was curious to see another person. He looked preoccupied. What a coincidence! It was Bruce Di Labio, a birding guide from my home town of Ottawa, playback machine in hand, trying to coax out a Yellow Rail after having had success with the Nelson's Sparrow. He was scouting good birds through southern Manitoba (prairie potholes, etc.) for his tour the next day. He told me the fascinating story of Yellow Rails calling along with the *tic-tic-*

François Shaffer

tic, tic-tic-tic of the night train traveling through Richmond Fen, south of Ottawa. And recently, Fred Helleiner (2013, 35) wrote of an incident from some decades ago when one bird "created a stir … when its repetitive ticking noises behind a gas station in northern Ontario became the subject of a police investigation—perhaps a time bomb?"

The Yellow Rail is rather difficult to see in its breeding range. In winter, watch for birds flushed by moving farm machinery in agricultural fields. This is best accomplished by joining events like Louisiana's "Yellow Rails and Rice Festival" and watching closely for white wing-patches.

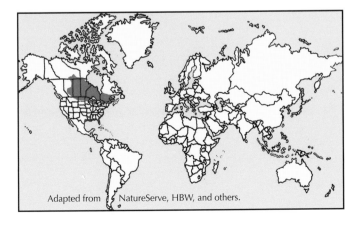

Adapted from NatureServe, HBW, and others.

NatureServe
Conservation Status

Global: ...... Apparently Secure

Canada: ..... Apparently Secure

US: .... Vulnerable

Mexico: ..... Possibly Extirpated

www.natureserve.org

Black Rail

Scientific: *Laterallus jamaicensis*
Français: Râle noir
México: Polluela Negra
Order: Gruiformes
Family: Rallidae

"The flight is much more feeble than that of any other rail with which I am familiar; the bird seemed barely able to sustain its weight in the air, while its legs dangled down helplesssly behind" (Griscom 1915, 228). The Black Rail's short flights are typical for rails. Like the Yellow Rail, the Black Rail's anatomical design favors running like a mouse through the fine-stemmed emergent vegetation (e.g., sedges, rushes, grasses) rather than flying with powerful wing-beats. It is thought to use the tunnels made in vegetation by meadow mice (*Microtus* spp.) Perhaps as a result of its weak flight, the Black Rail is particularly susceptible to avian predators such as herons during extreme high-tides when there is nowhere to go, stranded between water and land with little tidal-marsh remaining. Fortunately, however, site managers often attempt to leave some natural surrounding upland areas, which Black Rails can use for temporary escape.

The Black Rail shares the slightly larger Yellow Rail's predilection for shallow-water wetlands (squishy substrate with scattered small, shallow pools), and thus faces the same major threats to its habitat—drainage and degradation by humans. Having looked for this species in the Great Basin in 2014 during the current drought plaguing the region, and simply not finding any

alanmurphyphotography.com

habitat (i.e., any water) where such existed before, I am concerned about climate change and its projected more-intense dry spells in that region. For now, the Yellow Rail has the advantage of a relatively large breeding-stronghold in the less disturbed parts of central Canada (e.g., James Bay and Hudson Bay) compared to the Black Rail's scattered population distribution in the US. Nearly all US populations of the Black Rail likely declined greatly in the 1990s (Eddleman et al. 1994).

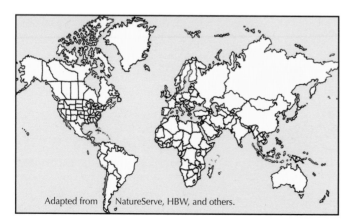

Adapted from NatureServe, HBW, and others.

NatureServe
Conservation Status

Global: ....Vulnerable

Canada: .... Not Applicable

US: ....Vulnerable

Mexico: ....Critically Imperiled

www.natureserve.org

ROBERT ALVO

Clapper Rail

Scientific: ***Rallus longirostris***
Français: Râle gris
México: Rascón Picudo
Order: Gruiformes
Family: Rallidae

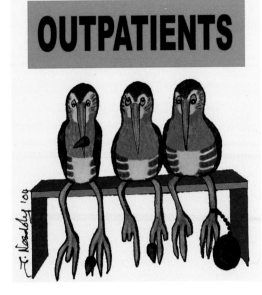

"Another enemy has arisen to make the life of this bird miserable—a certain mussel once imported from the East.... When our native son steps carelessly, it closes its doors with a bang—and often seizes the hapless rail by the toe. So common is this that many specimens with maimed feet or missing toes have been taken, and a few have been captured right where they were being held captive by the mussels. Others, more fortunate in escaping, are nevertheless condemned to drag about a ball on the foot—a mass of dried mud and trash of which the mussel is the unyielding nucleus" (Dawson 1923).

The Clapper Rail was split into three species in 2014: the Ridgway's Rail (*Rallus obsoletus*) of California and Arizona, the Clapper Rail (*R. crepitans*) of eastern North America, and the Mangrove Rail (*R. longirostris*) of South America (Chesser et al. 2014). Insufficient information is available at the time of writing to have separate pages for the "new" Ridgway's Rail and the "new" Clapper Rail. (As noted in the Introduction, this is the only American Ornithologists'

Bill Schmoker

Union change made in 2014 that is not adopted in this book.) Thus, the species names given at the top of this page and the conservation status ranks given below apply to the "old" Clapper Rail, whereas the cartoon and first paragraph apply to the "new" Ridgway's Rail. The photo is a "new" Clapper Rail (from Texas). The map, however, indicates the relative ranges of the three newly recognized species (Maley and Brumfield 2013).

This page serves as a useful reminder that taxonomy, especially in birds, is passing through the age of molecular genetics.

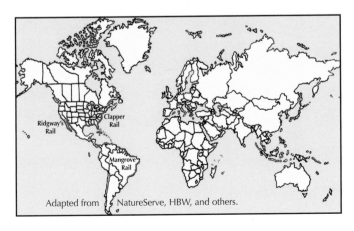

Ridgway's Rail

Clapper Rail

Mangrove Rail

Adapted from NatureServe, HBW, and others.

NatureServe
Conservation Status

Global: .... Secure

Canada: .... Not Applicable

US: .... Secure

Mexico: ..... Apparently Secure

www.natureserve.org

King Rail

Scientific: ***Rallus elegans***
Français: Râle élégant
México: Rascón Real
Order: Gruiformes
Family: Rallidae

Rails are a wonderful challenge to hunt by foot. Upon detecting hunters, rails "flatten and run, squirting through the marsh-grass like mercury" (Thompkins 2011). The hunters chase them by "slogging through boot-sucking mud", trying to flush them. Once scared up, the rails fly erratically for a short distance, dive back into the thick marsh vegetation, then run quickly, only rarely flushing again. This is excellent exercise in the September heat, but only 300–2000 Texans per season take advantage of the 70-day opportunity.

Rail hunting is much more popular on the Atlantic Coast, where it is done by boat. Tide levels vary much more there than on the Gulf Coast, such that high tides flood the marsh and push the birds onto slivers of high ground. One hunter poles the boat from the stern while the other shoots from the bow. (The internet has some fascinating videos of rail hunting.) Rails have a distinctive taste and are a treat when properly prepared (http://www.dto.com/hunting/).

Of our six rail species, three (Clapper Rail, Virginia Rail, and Sora) are globally Secure and hunted legally in parts of the US and Canada, whereas the other three are the cause of some global conservation concern (Black Rail: Vulnerable; Yellow Rail and King Rail: Apparently Secure). Black Rails are protected from hunting in the US and do not occur in Canada. Yellow Rails are protected in both Canada and the US. The King Rail is protected in Canada, but despite its alarming decline in numbers over the past decades, hunting is still permitted in all southern coastal US states from Texas to Delaware, and in Connecticut. Its status as a hunted species in the US is therefore being reassessed through an ongoing conservation plan (Cooper 2007).

In places where muskrats are trapped, King Rails get killed because they use the same runways. One muskrat trapper in Maryland caught 50 of these birds in one season (Poole et al. 2005).

Noppadol Paothong

Courtship is rarely seen, much less photographed, in this secretive species.

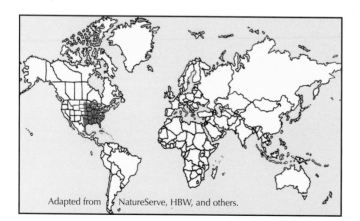

Adapted from NatureServe, HBW, and others.

NatureServe
Conservation Status

Global: ...... Apparently Secure

Canada: ....Imperiled

US: .... Apparently Secure

Mexico: ..... Apparently Secure

www.natureserve.org

Virginia Rail

Scientific: ***Rallus limicola***
Français: Râle de Virginie
México: Rascón Limícola
Order: Gruiformes
Family: Rallidae

The expression "skinny as a rail" refers equally well to the rails of a railway track as to the bird. Moving through emergent vegetation, the bird compresses its wings and plumage closely against its laterally-flattened body. Rails have very high leg-muscle to flight-muscle ratios, making them excellent runners but poor fliers. They usually run from danger secretively rather than fly away conspicuously.

Flexible vertebrae make it easier for Virginia Rails to maneuver through vegetation, and abrasion of the head feathers is reduced by special feather tips. Valves of mucus-covered skin protect the inner membrane of nasal passages from pointy vegetation tips (Conway 1995).

Virginia Rails and Soras often occur in the same marshes. The longer-billed Virginia Rails feed mostly on invertebrates (especially by probing the soft ground and litter), whereas Soras are omnivorous, pecking wetland-plant seeds and invertebrates. To compensate for their difficulty in providing their young with enough protein-laden invertebrates for rapid growth, Soras compete with Virginia Rails by defending territories and using their black faces to enhance threat displays. Soras also allow themselves more time to raise all their chicks by staggering the hatching period (through staggered egg-laying and incubation) and feeding only the younger ones. Virginia Rails have fewer threat displays and are less aggressive (Kaufmann 1989).

Dan Parent

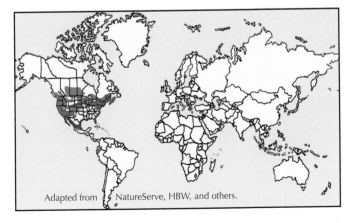

Adapted from NatureServe, HBW, and others.

NatureServe
Conservation Status

Global: .... Secure

Canada: .... Secure

US: .... Secure

Mexico: ..... Apparently Secure

www.natureserve.org

Sora

Scientific: *Porzana carolina*
Français: Marouette de Caroline
México: Polluela Sora
Order: Gruiformes
Family: Rallidae

"The Sora is unquestionably *the rail* of North America. It is the most widely distributed and the best known of its tribe. Throughout its wide breeding range its cries are among the most characteristic voices of the marshes" (Bent 1926, 303). Not surprisingly, it is also considered to be the most abundant of our six rail species.

Most rails inhabit dense cover, in which their long toes and light weight allow them to walk on floating or emergent vegetation and on mud. Individuals usually cannot see each other between territories, so they vocalize to communicate. Soras call frequently during spring migration and the breeding season. Their most common vocalization is a descending call reminiscent of a horse's whinny. Like other rails, Soras often vocalize at night. Interestingly, Soras will call in response to loud noises, even the slamming of a car door.

Rails are weak fliers and tend to migrate at low altitudes, yet they can migrate long distances. The entire family is renowned for its ability to colonize remote oceanic islands, and this is partly a result of the ease with which rails are blown off course by unfavorable winds. Two indications that Soras migrate mostly at night are that individuals have been reported to call at night in flight during spring migration, and that Soras are frequently found dead near the base of lighted towers (Melvin and Gibbs 2012).

Although most rail species live in wetlands, their evolutionary origin is not necessarily aquatic. In fact the opposite seems to be true: the most primitive living rail, the Nkulengu Rail (*Himantornis haematopus*) of central Africa, is a forest bird, as are other primitive or unspecialized species. In contrast, most of the recently differentiated and specialized species are marsh-dwellers or aquatic (Taylor 1996).

Dennis Paulson

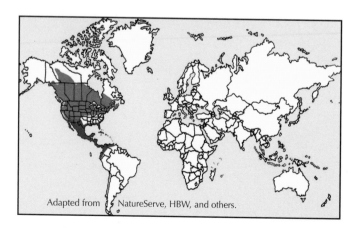

Adapted from NatureServe, HBW, and others.

NatureServe
Conservation Status

Global: .... Secure

Canada: .... Secure

US: .... Secure

Mexico: ..... Apparently Secure

www.natureserve.org

Purple Gallinule

Scientific: *Porphyrio martinicus*
Français: Talève violacée
México: Gallineta Morada
Order: Gruiformes
Family: Rallidae

Cooperative breeding is a social system in which individuals (i.e., "helpers") provide parental care to young that are not their own. It occurs in about 9% of the world's birds (including Purple Gallinules), 2% of mammals, less than 1% of fishes, less than 0.1% of insects, and only a few species of arachnids and crustaceans. The biological parents and their helpers reap benefits and suffer costs that have been measured in numerous studies attempting to explain why cooperative breeding occurs and how it evolved (Eggert 2014).

Purple Gallinules migrate from the tropics to breed in temperate zones in both the Northern and Southern Hemispheres. Other populations are resident, and such pairs can raise more than one brood a year. Young from one brood may feed those of the next brood, and even three broods can be involved. Families also include other possibly-related adults that feed and defend the chicks. Similar cooperative breeding behavior is known in at least seven other rallid species. Breeders benefit by raising more young and making fewer chick-feeding visits, whereas helpers benefit by increasing the production of young that are often their relatives. Helpers also gain experience in raising chicks, avoiding predators, finding food, and defending territories.

Because they feed on Water Hyacinth (*Eichhornia crassipes*) and Hydrilla (*Hydrilla verticillata*), two invasive introduced plants, Purple Gallinules tolerate and perhaps benefit from the infestation of these weeds caused by fertilizer runoff into wetlands.

Robert Alvo

Great losses of wetlands from the 1950s to the 1970s in much of this species' US range was offset to some degree by the creation of rice fields, national wildlife refuges, and water-conservation impoundments. However, the trend toward growing rapidly-maturing varieties of rice is problematic because harvesting often occurs before the chicks have had the opportunity to grow sufficiently to avoid the machinery (West and Hess 2002).

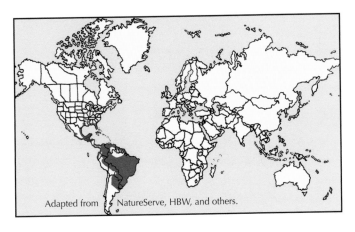

Adapted from NatureServe, HBW, and others.

NatureServe
Conservation Status

Global: .... Secure

Canada: .... Not Applicable

US: .... Secure

Mexico: .... Secure

www.natureserve.org

Common Gallinule

Scientific: *Gallinula galeata*
Français: Gallinule d'Amérique
México: Gallineta Frente Roja
Order: Gruiformes
Family: Rallidae

Lake shorelines have platforms made by beavers or nesting loons (see first photo, p. 78), and flattened areas where otters or herons have fed. In marshes, however, we find platforms of aquatic vegetation made by muskrats, rice rats, gallinules, coots, terns, and grebes. Common Gallinules, for example, construct several kinds of platforms that can be used to lay eggs, brood chicks, or for displaying. One pair may build numerous platforms. Muskrats help gallinules and other birds not only by removing vegetation from some areas for their lodge clearings, canals, and runways, but also by building feeding and nesting platforms that are often used later by the birds. I have not seen a field guide to the platforms that aquatic birds and mammals make and/or use. Such a book could be enhanced by including artificial platforms made by humans to improve bird-nesting opportunities.

The Common Gallinule was considered the same species as the Old World's Common Moorhen, but significant differences in genetics, vocalizations, and bill and shield morphology justified separation into two species. The shield, lying above and meeting the bill, is rounded at the top and widest in the middle in Old World birds, but squared off and widest near the top in New World birds. More biological research has been done on Old World birds than on New World ones. Further research is required on New World birds to determine which findings apply to both species.

The Common Gallinule (a) and the Common Moorhen (b and c) look identical except for a small difference in the shape of the shield, not readily visible in these photos (see text).

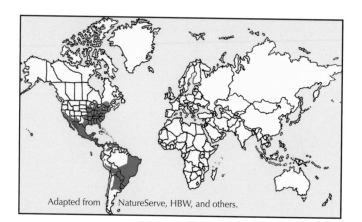

Adapted from NatureServe, HBW, and others.

NatureServe
Conservation Status

Global: .... Secure
Canada: ..... Apparently Secure
US: .... Secure
Mexico: .... Secure

www.natureserve.org

American Coot

Scientific: *Fulica americana*
Français: Foulque d'Amérique
México: Gallareta Americana
Order: Gruiformes
Family: Rallidae

I **TOLD** you, only leave the hairdye on for 10 minutes !!!!!

If the male King Eider is the bird that aliens would snatch first (see p. 42), the young American Coot chick is the one that would remind them most of home. Its striking head-color stimulates parents to feed it, as shown in scientific experiments.

The strange-looking chicks are symbolic of other odd stories associated with the species. American Coots are assumed to usually follow aquatic habitats during migration, but they sometimes cross large regions of inappropriate habitat such as forests with no wetlands nearby, and open ocean many kilometers off the North American coast. One migrating bird flew through an open window into the opposite wall of a top-floor hotel room. And a long flock of 10,000 coots was observed over three spring days moving north *on foot* (Prill 1931).

First Nations people used to undertake communal hunting drives to net or hand-capture thousands of coots for food, and this species is still popular with hunters, although less so than waterfowl (ducks, geese, and swans). When thousands of coots were present during an outbreak of avian cholera in Virginia in 1975, thus posing a threat to waterfowl, 6000 coots were sprayed with the wetting agent "PA-14" and killed to reduce the threat.

Coot nests are floating structures of aquatic vegetation. Nest-cup depth and rim height above water decrease with time as the structure becomes waterlogged. This makes nest-measuring tricky. Nest vegetation is in constant decay, another reason why the parents must continually add vegetation. Nests containing 16–22 eggs may be the result of laying by more than one female, as often occurs in other birds, but alternatively it may indicate that the female has laid a second clutch without removing the remaining eggs from a clutch that had been partially depredated. Another species oddity is the variability in the start of incubation relative to the number of eggs already in the nest, combined with a long period of staggered hatching—a precise incubation period is thus very difficult to determine (Brisbin and Mowbray 2002).

Robert Alvo

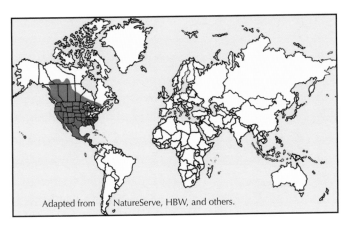

Adapted from NatureServe, HBW, and others.

NatureServe
Conservation Status

Global: .... Secure
Canada: .... Secure
US: .... Secure
Mexico: .... Secure

www.natureserve.org

Limpkin

Scientific: *Aramus guarauna*
Français: Courlan brun
México: Carrao
Order: Gruiformes
Family: Aramidae

The Limpkin's story, like that of the Snail Kite, is set in Florida and centered on an apple snail (*Pomacea paludosa*) on which both birds depend. South of Florida, where this snail species does not occur, other apple snails (*Pomacea spp.*) are eaten by both birds. The Limpkin, or "crying bird", is "the voice of one [person] crying in the wilderness" (Bent 1926, 254). We are fortunate to be able to hear this "most disagreeable note of any of our native birds" (Bryant 1861); so many tasty Limpkins were easily killed for food in the past that Bent predicted the species' demise in Florida.

"It is easy to detect the presence of Limpkins by looking for the deposits of the empty shells of apple snails. The birds have favorite feeding places where they bring the snails; one can often find a number of empty shells around some old log or snag or on an open place on a bank"

Dennis Paulson

(Bent 1926, 257). These sites are analogous to the feeding places of otters and herons mentioned in the Common Gallinule account (p. 160), and may warrant mention in the "platform field guide" proposed. Limpkin feeding places are called "extraction sites" because they serve as solid bases where the birds can deftly extract snails from their shells. Florida apple snails live only about 12–18 months and die within a few weeks of reproducing, so populations must have some breeding success each season to remain viable. One water drawdown by humans can thus be devastating (Bryan 2002).

Limpkins eat rotten wood, usually before or after foraging for mollusks, and it is sometimes offered in courtship feeding. Perhaps these freshwater birds find sodium in rotten wood, as is the case in Mountain Gorillas (*Gorilla beringei beringei*) (Than 2006).

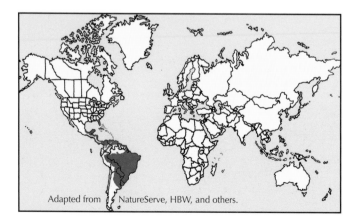

Adapted from NatureServe, HBW, and others.

NatureServe
Conservation Status

Global: .... Secure

Canada: .... Not Applicable

US: .... Vulnerable

Mexico: .... Secure

www.natureserve.org

Sandhill Crane

Scientific: *Grus canadensis*
Français: Grue du Canada
México: Grulla Gris
Order: Gruiformes
Family: Gruidae

"His great stature gives him the range almost of that of a man; his eye is wondrously keen, telescopically so; it is so near the top of his head that he can peer over the crest of a knoll and see without being seen" (Laing 1915). "When only a slight wind is blowing, these rich, bugle-like notes can be heard farther than the bird can be seen. Several times I have examined, for some moments in vain, the horizon before the authors sailed in view" (Visher 1910, 115).

The Sandhill Crane is the world's most abundant crane. Its numbers seem to be stable or increasing range-wide, except in the nonmigratory Florida population. Yet of all North America's game birds, this species produces the fewest young per pair annually (mostly because of high nest-predation rates), and birds start breeding only when two to eight years old.

When you see a group of Sandhill Cranes, it may consist of breeding pairs, families, paired nonbreeders, groups of unpaired nonbreeders, or a mixture of all of these. Families typically include a breeding pair, their youngest offspring, often their two-year-olds, and sometimes much older offspring. Good luck sexing them: even if you could sneak up to the birds and examine their cloacas, you wouldn't be able to tell whether they were females or males. If the breeding pair performs their Unison Call, however, you can distinguish them by the male's more drawn-out and lower-pitched part of the duet (Gerber et al. 2014).

Historically in many parts of the world, including among various North American First Nations, cranes have been symbolic of good things: longevity, fidelity, health, good fortune, happiness, peace, and even authority. In North America, especially in the US, crane festivals now occur annually during the nonbreeding season.

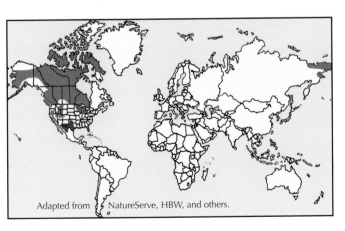

Larry Masterr

Adapted from NatureServe, HBW, and others.

NatureServe
Conservation Status

Global: .... Secure
Canada: .... Secure
US: .... Secure
Mexico: .... Not Applicable

www.natureserve.org

Whooping Crane

Scientific: *Grus americana*
Français: Grue blanche
México: Grulla Blanca
Order: Gruiformes
Family: Gruidae

The male Whooping Crane, which stands almost at a whopping 1.5 meters, is North America's tallest bird. This species stands tall also as an international conservation success story that is still in the making.

Whooping Cranes are thought to have numbered more than 10,000 birds when they bred throughout the prairies before European colonization. They wintered from New Jersey south to Florida and along the Gulf Coast through Texas to northeastern Mexico. Overhunting and the loss of habitat almost caused the species' extinction, such that by 1860 only an estimated 1400 individuals survived. The all-time low of 15 birds was reached in the 1940s. Fortunately the Aransas National Wildlife Refuge was created in Texas in 1937, followed by the discovery of the breeding grounds in 1954 at Wood Buffalo National Park along Alberta's northern boundary with the Northwest Territories. In the 1950s and 1960s, hunters were educated to prevent unintentional shooting of Whooping Cranes during migration. In 1967 a captive-breeding program was initiated in Maryland, where several hundred eggs have since hatched in captivity or in Sandhill Crane nests. Several other captive flocks have also

Juan Bahamon

been established, for example at the International Crane Foundation in Wisconsin and at the Calgary Zoo, for producing young birds that have been introduced in Wisconsin (a migratory flock) and in Florida (a nonmigratory flock). Another conservation program involves using ultralight aircraft to teach young birds where to migrate.

A total of about 600 Whooping Cranes now exist, with approximately 400 wild birds and 200 in captivity, but only the Aransas-Wood Buffalo population is wild, self-sustaining, and migratory.

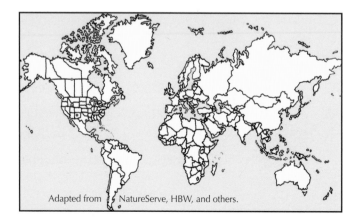

Adapted from NatureServe, HBW, and others.

NatureServe
Conservation Status

Global: ..... Critically Imperiled

Canada: .... Critically Imperiled

US: .... Critically Imperiled

Mexico: ... Presumed Extirpated

www.natureserve.org

Black-necked Stilt

Scientific: *Himantopus mexicanus*
Français:　Échasse d'Amérique
México:　　Monjita Americana
Order:　　Charadriiformes
Family:　　Recurvirostridae

Belly-soaking is one of several methods that shorebirds breeding in hot environments can use to prevent their eggs from overheating (see Wilson's Plover account, p. 173). But other bird groups do it too, many wetting their feathers in different ways and for different reasons. Charadriiformes (shorebirds, gulls, terns, and their allies) use it the most. However, Purple Martins have been observed belly-soaking too, in artificial nesting situations (so-called "Purple Martin apartments") where nests may get hotter than in natural nesting habitats. While some birds such as Black-necked Stilts wet their ventral feathers by standing in water and bending their knees so those feathers come into contact with the water, sometimes repeatedly, terns skim through the water or plunge-dive to wet the feathers. Other birds run through the water with their wings spread. Contact between the belly feathers and the water may take only a few seconds or last up to 10 minutes. The function of belly-soaking may be to cool the eggs, the chicks, or the adults, or to humidify the nest. Parents in some species puff out their wet breast feathers for the chicks to drink from. In Black-necked Stilts, one pair may make more than 100 trips to water in one day. A literature review and further study of belly-soaking would be useful because it is a sign of heat stress in a world that is experiencing more extreme weather in this period of rapid climate change.

Juan Bahamon

"At times they seem a bit wabbly on their absurdly long and slender legs.... The legs are so long that when the bird is feeding on land it is necessary to bend the legs backward to enable the bill to reach the ground" (Bent 1927, 52).

Adapted from NatureServe, HBW, and others.

NatureServe
Conservation Status

Global: .... Secure
Canada: .... Imperiled
US: .... Secure
Mexico: .... Secure

www.natureserve.org

American Avocet

Scientific: *Recurvirostra americana*
Français: Avocette d'Amérique
México: Avoceta Americana
Order: Charadriiformes
Family: Recurvirostridae

"The use of the avocet's recurved bill is clearly explained by the manner in which the bird procures its food. In feeding they wade into the water and drop the bill below the surface until the convexity of the maxilla probably touches the bottom. In this position they move forward at a half run and with every step the bill is swung from side to side sweeping though an arc of about 50° in search of shells and other small aquatic animals. The mandibles are slightly opened, and at times the birds pause to swallow their prey. It is evident that birds with a straight or a downward curved bill could not adopt this method of feeding" (Chapman 1891, 321) now appropriately called "scything".

The world's four avocets comprise the genus *Recurvirostra*. Our species is unique among them in having distinct breeding and wintering plumages. Before learning to fly, large young have a plumage color and pattern that makes them look like adults. (Even the downy chick's head and neck are orange.) Large young tend to run from predators rather than hide, and use adult-like foraging behaviors like scything. This "auto-mimicry", or mimicry within the species, is thought to help protect the young from predators that confuse them with adults, which can fly. You can pick out the juveniles by their slightly smaller size, fluffier overall look, and more swollen ankles (Sibley 2001).

Edgar T. Jones

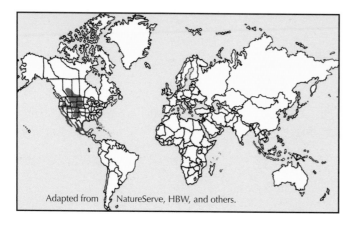

Christian Artuso

Adapted from NatureServe, HBW, and others.

NatureServe
Conservation Status

Global: .... Secure

Canada: .... Secure

US: .... Secure

Mexico: .... Secure

www.natureserve.org

American Oystercatcher

Scientific: *Haematopus palliatus*
Français: Huîtrier d'Amérique
México: Ostrero Americano
Order: Charadriiformes
Family: Haematopodidae

Most of the world's 11 oystercatcher species, all comprising the genus *Haematopus*, breed along marine coasts whose habitats range from rocky shores to beaches of shingle and sand to coastal marshes. They feed mostly on bivalves, gastropods, and marine (polychaete) worms. Their most characteristic morphological feature is the long, red, heavy bill, which they use to pry open or hammer open bivalves whose two valves (half-shells) are closed, or to tap off or lever off chitons and limpets, which are not bivalves, from rock. With coiled snails, soft tissue must be pulled out or the shell must be broken, always using the ingenious tool. Attack success depends on the size of the shellfish, the strength of its shell, and the ability of the oystercatcher to manipulate it or remove it from its substrate. Ornithological economists can calculate how small a particular prey species can be before it takes the bird too much energy to obtain the diminishing return in calories (Hockey and Bonan 2013).

While some oystercatcher species breed inland, the American Oystercatcher is strictly marine. Its species name *palliatus* means cloaked, referring to its black breast, neck, head, and back, and is reminiscent of one of J. R. R. Tolkien's "black riders" in *The Lord of the Rings*. Its feeding tool is laterally compressed like the double-edged knife used to shuck oysters, and the chisel-like tip is perfect for extracting shellfish from rocks. While the bird is wading through shallowly submerged mussel or oyster beds, the trick is to find one whose valves are open and

David Laliberte

thrust the bill through the open valves, cutting the muscle that connects them, and finally extracting the soft parts of the body. Another foraging method is to remove a bivalve from the water, place it on a surface, line it up, and hammer it at the point where the muscle attaches the two valves. Once the shell is broken, the bird cuts the muscle. One of the dangers for foraging oystercatchers lies in getting the bill clamped by a stubborn bivalve and then drowning when the tide rises (American Oystercatcher Working Group et al. 2012).

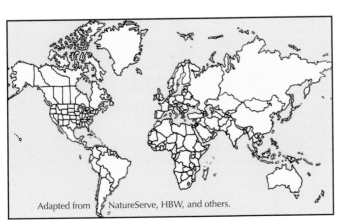

Adapted from NatureServe, HBW, and others.

NatureServe
Conservation Status

Global: .... Secure
Canada: .... Not Applicable
US: .... Secure
Mexico: .... Secure

www.natureserve.org

Black Oystercatcher

Scientific: *Haematopus bachmani*
Français: Huîtrier de Bachman
México: Ostrero Negro
Order: Charadriiformes
Family: Haematopodidae

Oystercatchers nest above the high tide line, but feed in the intertidal zone mainly during low tide when their invertebrate food is most accessible. The Black Oystercatcher prefers rocky shores with considerable surf action, where it walks through the shallow water visually searching for prey that are similar to that of the American Oystercatcher (mostly shellfish). Waves vary in height and intensity, so the bird must always be ready to jump straight up to avoid losing control of its position. Alternatively, it can move to higher ground before returning to resume foraging. Theoretically, the available foraging time depends on the length of time that favorable conditions are allowed by the tides, yet Black Oystercatchers often suspend their activities during low tide to give themselves enough time to digest (Andres and Falxa 1995).

In altricial birds the defenseless young are fed entirely by their parents until after they leave the nest, whereas in precocial birds the young are mobile and feed themselves. Between these two extremes is a gradient somewhere along which all bird species lie. Oystercatchers are unique in that they are mobile and abandon the nest within hours of hatching, yet are

Bill Schmoker

fed entirely by their parents until well after the young can fly. This allows them to minimize the risk of predation by being able to run away or hide, while at the same time maximizing their growth rate by having their experienced parents forage for them. In this apparent "super strategy", the only down-side that has been suggested is that the parents, both of which feed the chicks, are often energetically stressed during the chick-rearing period. In the Black Oystercatcher and three other oystercatcher species, some of the chicks may starve. When food is limited, the sibling hierarchy in this species inevitably leads to the dominant chick surviving at the expense of its one or two siblings (Hockey and Bonan 2013).

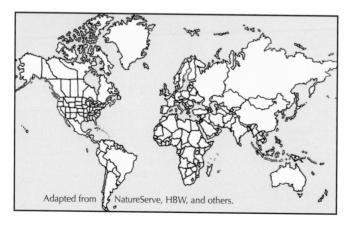

Adapted from NatureServe, HBW, and others.

NatureServe
Conservation Status

Global: .... Secure

Canada: ..... Apparently Secure

US: .... Secure

Mexico: ..... Apparently Secure

www.natureserve.org

Black-bellied Plover

Scientific: *Pluvialis squatarola*
Français: Pluvier argenté
México: Chorlo Gris
Order: Charadriidae
Family: Charadriiformes

Robert Elsmore

The Black-bellied Plover is "an aristocrat among shorebirds, the largest and strongest of the plovers,… the athlete of the wild birds of the North,… a leader of its tribe,… distinguished-looking in its handsome spring livery of black and white…, and always dignified and imposing…. It is as a game bird that … [it] has achieved its greatest reputation,… for it is not only a large plump bird but it is a swift flier, and one of the wariest, most sagacious, and most difficult of the beach birds [shorebirds] to secure…. The sportsman [hunter] must be familiar with its habits in the locality where he is shooting, must be well concealed in a skillfully made blind, and must know how to imitate its notes perfectly. The old birds are particularly wary and … succumb only to those sportsmen who have served a long apprenticeship" (Brandt, H. W., Mackay, G. H., and Bent, A. C., in Bent 1929, 154–165). OK, OK, but Bent (1929, 154) exaggerated when he suggested that "the lesser fowl … look to it for leadership".

Its unique black "armpits" (axillars) are the most reliable field mark in all plumages. You can see them on birds flying far away. This key field mark conjures up a different image for me than Bent's (1929) aristocrat: namely an arm-chair athlete with hairy armpits, drinking beer, and watching wrestling on television. It's your decision as to whether you prefer to remember this species as an aristocrat or a bum, be you a crusty old "twitcher" from the Old World, where it is known as the Grey Plover, or a neophyte North American birder.

David Laliberte

Peter Sproule

Larry Master

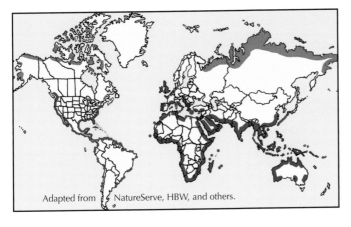

Adapted from NatureServe, HBW, and others.

NatureServe
Conservation Status

Global:........................ Secure
Canada:................. Vulnerable
US:............................. Secure
Mexico:.......... Not Applicable

www.natureserve.org

American Golden-Plover

Scientific: *Pluvialis dominica*
Français: Pluvier bronzé
México: Chorlo Dorado Americano
Order: Charadriiformes
Family: Charadriidae

The American Golden-Plover is one of the migrant champions of the bird world in terms of flight speed (up to 180 km/hour) and distance traveled. After breeding in the high Arctic tundra, it migrates 13,000 km to the pampas of South America via the western coasts of Hudson Bay and James Bay (a critical staging region), the Canadian Maritimes, then over the open western Atlantic Ocean to its destination. In spring it completes its elliptical annual migration by passing over the interior of North America. Weather permitting, the autumn portion from James Bay to South America is often accomplished in a nonstop flight. The elliptical route allows the American Golden-Plover (also the Hudsonian Godwit and Buff-breasted Sandpiper) to take advantage of favorable winds at different times of year.

Long migrations require huge amounts of energy, so American Golden-Plovers gorge themselves for two months to build a sufficient storage of body fat, which is the best-known fuel for migrating birds. Fat can be packed under the skin, around organs, and in numerous body cavities such that its weight is well-distributed. (Proper weight distribution is critical for efficient, controlled flight in birds, as in aircraft.) Metabolizing fat yields more than twice the amount of energy as does the burning of carbohydrates or protein. In addition, more than twice the amount of water is released when fat is burned in comparison with carbohydrates and protein, and this is critical to prevent dehydration. Fat may represent as much as 50% of the body weight of American Golden-Plovers that are ready to migrate. This is why gunners turned their sights on this species and on another fattened long-distance migrant, the Eskimo Curlew, when Passenger Pigeon numbers were in decline as early as 1860—migration-primed birds made for excellent meals (Sibley 2001; Podulka 2004; Johnson and Connors 2010).

Edgar T. Jones

Larry Master

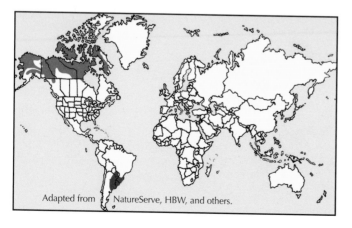

Adapted from NatureServe, HBW, and others.

NatureServe
Conservation Status

Global: .... Secure

Canada: ..... Apparently Secure

US: .... Secure

Mexico: .... Not Applicable

www.natureserve.org

ROBERT ALVO

Pacific Golden-Plover

Scientific: *Pluvialis fulva*
Français: Pluvier fauve
México: Chorlo Dorado del Pacífico
Order: Charadriiformes
Family: Charadriidae

The name "golden-plover", forming part of the English name for three of the world's shorebird species (American, Pacific, and European Golden-Plovers), applies well to all three species in any plumage, for they always show some gold or yellow on their upper side. "The downy chicks are among the loveliest of all young birds, their yellow backs being startlingly different from the usual blacks, browns, and grays affected by most newly hatched youngsters of the shorebird clan" (Gabrielson and Lincoln 1959). The only other *Pluvialis* species, the Black-bellied Plover, also shows gold or yellow in every plumage (including in hatchlings) except in breeding adults, which are black, white, and gray (Cramp 1983).

Whereas scolopacids (sandpipers and their allies) use their sense of touch (tactile probing) to forage and feed, plovers rely more on their eyes. Indeed, plovers have larger optic brain lobes than do scolopacids. Some plovers regularly feed at night, and this habit may be related to the high density of rod cells in the retinas of plovers relative to scolopacids. Rod cells are sensitive to low light, thus allowing these birds to feed in darkness. Pacific Golden-Plovers have been seen night-feeding, but it is not clear how often they do so. Possible reasons for night-feeding include a reduction in competition for food with other species, and the avoidance of diurnal predators. Or, maybe night-feeding increases the chances of finding certain prey. Foraging into the night in addition to during the day may be required for the birds to meet their daily food requirements. All these reasons are plausible enough that each one may be involved, perhaps changing in importance under different situations.

Netta Smith

Foraging plovers often raise one leg and shake it rapidly to disturb the wet substrate, which causes small invertebrates to move and become easier to see and catch. In grassland situations, foot-shaking and foot-tapping sound like rain or burrowing predators; these movements cause some prey, such as earthworms, to rise to the surface where they are easily seen and captured (Sibley 2001; Johnson and Connors 2010).

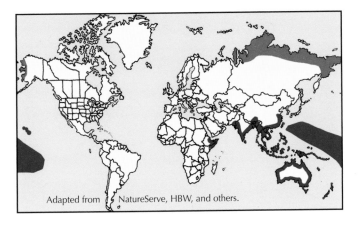

Adapted from NatureServe, HBW, and others.

NatureServe
Conservation Status

Global: .... Secure

Canada: .... Not Applicable

US: .... Secure

Mexico: .... Not Applicable

www.natureserve.org

Snowy Plover

Scientific: *Charadrius nivosus*
Français: Pluvier neigeux
México: Chorlo Nevado
Order: Charadriiformes
Family: Charadriidae

"HONEY, I'VE ELOPED"

Until recently, the New World's Snowy Plover was considered the same species as the Old World's Kentish Plover, whose species name *Charadrius alexandrinus* refers to Alexandria, Egypt, where Carl Linnaeus's type specimen came from. Separation of the two species is based on differences in morphology, genetics, and male vocalizations. Both species at times use an unusual polygamous breeding system in which one member of the pair, usually the female, deserts the other parent and the brood to nest again with a new mate during the same breeding season. The deserting parent sometimes travels hundreds of kilometers. On the California coast, for example, female and male Snowy Plovers move up to 660 and 840 km, respectively, to renest. In both species the female does most of the daytime incubation, whereas the male takes the night shift. Females may need to forage at night to regain energy lost to laying eggs, and their dull coloration compared to that of males makes them less visible on the nest to daytime predators. Being more aggressive than females, males may be better at defending the nest from predators at night, and their brighter coloration likely makes them better at defending breeding territories from other Snowy Plovers during the day.

Snowy Plovers nest on the ground on beaches and similar shoreline habitats, so their nests are exposed to wind, humans on foot, and vehicles. These places are also in great demand for human development—hence this widespread species' Vulnerable global status (Page et al. 2009).

Edgar T. Jones

Jenny Erbes

Adapted from NatureServe, HBW, and others.

NatureServe
Conservation Status

Global: .... Vulnerable

Canada: .... Critically Imperiled

US: .... Vulnerable

Mexico: .... Vulnerable

www.natureserve.org

Wilson's Plover

Scientific: *Charadrius wilsonia*
Français: Pluvier de Wilson
México: Chorlo Pico Grueso
Order: Charadriiformes
Family: Charadriidae

The Wilson's Plover is a coastal species throughout the year, only occasionally occurring far inland. It eats crustaceans and some insects, which it finds mostly in intertidal areas and above the high-tide line on sandy beaches. Most of its diet consists of fiddler crabs (*Uca* spp.), up to 98% in a study in Venezuela. This group of crabs, of which there are roughly 100 species in the world, have two claws as do other crabs, but in male fiddler crabs one of the claws is fiddle-shaped and much bigger than the other. Males "play the fiddle" by moving their smaller claw (the bow) from ground to mouth during feeding. Fiddler crabs are also dubbed "calling crabs" because they communicate with each other using waves and gestures. The fiddling, waving, and gesturing are *not* known to be used for distracting predators such as Wilson's Plovers who are ready to pounce on them (see cartoon).

The open habitats above the high-tide line where Wilson's Plovers nest are subject to large temperature differences, and the parents adjust their behavior to keep the eggs from getting too hot or too cold. They sit on them to warm or cool them depending on the ambient air temperature. Parents can also shade the eggs from the sun. In very hot conditions they use belly-soaking, wherein they sit or crouch over the eggs with muddy or wet breast feathers. In a Texas study the highest nest temperature recorded was 48.5°C. In that nest located on pavement, overheating of the eggs (egg temperature 42.0°C) would have occurred when the nest temperature reached 43.2°C, but the parents prevented this by using belly-soaking. These three

Edgar T. Jones

methods of regulating egg temperature are used by other shorebirds as well. Belly-soaking is a useful method for cooling eggs, but it depends on a nearby water source and frequent trips by the parents, so it is not always feasible. Nesting too close to tidally-influenced water can lead to nest flooding.

As for the chicks, they may abandon the nest on foot 1–2 hours after the last chick hatches. The parents often shade them during the hottest time of day, but the chicks can also hide in vegetation (Corbat and Bergstrom 2000).

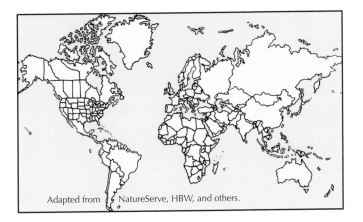

Adapted from NatureServe, HBW, and others.

NatureServe
Conservation Status

Global: .... Secure

Canada: .... Not Applicable

US: .... Apparently Secure

Mexico: .... Secure

www.natureserve.org

Common Ringed Plover

Scientific: *Charadrius hiaticula*
Français: Pluvier grand-gravelot
México: Chorlo Anillado Mayor
Order: Charadriiformes
Family: Charadriidae

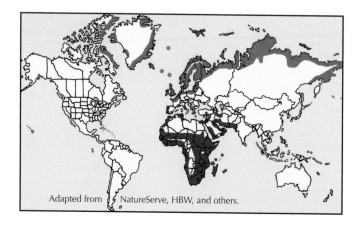

Climate change is affecting nature in many ways. Birds, with their gift of flight, are no less susceptible, even wide-ranging ones like Common Ringed Plovers. Changes in temperature, precipitation, and moisture are already affecting birds; climate is now more variable and weather more extreme. Habitats are changing (e.g., permafrost is melting).

Until recently, the natural balance was self-sustaining. Natural events usually occurred within a predictable time, and species lived in ecosystems that were generally predictable when viewed over the long term. Nature is becoming unbalanced in time and place. For some birds, life cycle events (e.g., feeding young birds) are no longer properly timed with other events such as the availability of food. The ranges of species that compose a given ecosystem do not all shift toward the North and South Poles at the same rate, so the ecosystems cannot remain intact under the current rate of climate change. Birds now have to contend with competitors, predators, and diseases to which evolution has not had sufficient time to help them adapt.

Picture the protected natural areas on the planet, then move each bird's range pole-ward, keeping in mind the patchy nature of most ranges. Keep going. At some point, many species will find themselves outside these favorable sites. Especially vulnerable bird groups include island or montane species, those that live in the Arctic or Antarctic, wetland species, and those of arid habitats. Seabirds and migrants are also very vulnerable. Species with poor conservation status or low dispersal ability, and those that live in patchy or uncommon habitats, are expected to be most

Mark J. Rauzon

adversely affected by climate change. After all those, what's left? Contractions of bird ranges should outnumber range expansions because species with low adaptability or a low dispersal ability will be pushed to places with no suitable habitats (Wormworth and Mallon 2006).

Common Ringed Plovers generally nest close to the high-water mark, which leaves them vulnerable to flooding during exceptionally high tides. Over the long term, rising sea levels are eroding low-lying coastal habitats, which harbor various species of birds.

Adapted from NatureServe, HBW, and others.

NatureServe
Conservation Status

Global: .... Secure

Canada: ..... Apparently Secure

US: .... Not Applicable

Mexico: .... Not Applicable

www.natureserve.org

Semipalmated Plover

Scientific: *Charadrius semipalmatus*
Français: Pluvier semipalmé
México: Chorlo Semipalmeado
Order: Charadriiformes
Family: Charadriidae

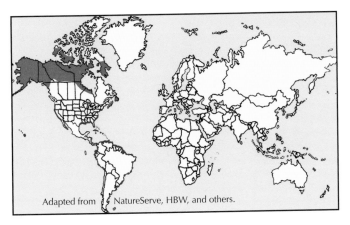

The Semipalmated Plover (SP) and the Common Ringed Plover (CRP) used to be treated as one species, but are easily separated by their calls. The CRP's breast band is slightly broader, its white eyebrow is more distinct, the gold eye ring is partial or absent, and its bill is slightly longer with more orange at the base. "Semipalmated" means "half-hand" and refers to birds with toes joined by a web only part way down. Two species with such toes are the SP and the Semipalmated *Sandpiper*. The CRP does not have this webbing that the SP shows between the basal part of the inner and middle toes, but both species have the webbing between the basal part of the middle and outer toes. These minor differences can be seen in the field at very close range. It is unlikely that they are related to major differences in habitat use by the two species (but see cartoon).

When building the nest, the male SP scrapes the ground with its breast while kicking its legs out behind to shape the nest cup. Walking away, it lines the nest by tossing material into it from up to one meter away. Both females and males throw nest material and peck at the sides of the nest throughout incubation. A number of bird species have been observed tossing nest material, but the reason for this behavior is unclear.

The SP is among the few plover species whose population sizes seem to be stable. This may be due to its generalist nature regarding food and habitat, or its widespread coastal wintering range. Also, suitable breeding habitat seems to be created by humans and breeding geese.

Larry Master

Gord Belyea

Feather lice (order Phthiraptera) are common on SPs in Churchill, Manitoba, and of the five shorebird species examined there, SPs contained the most internal parasites both in terms of the number of individuals and the number of species. It is not known why this is so (Nol and Blanken 2014).

Adapted from NatureServe, HBW, and others.

NatureServe
Conservation Status

Global: .... Secure

Canada: .... Secure

US: .... Secure

Mexico: .... Not Applicable

www.natureserve.org

Piping Plover

Scientific: *Charadrius melodus*
Français: Pluvier siffleur
México: Chorlo Chiflador
Order: Charadriiformes
Family: Charadriidae

The species name *melodus* means to sing pleasantly. Between the Piping Plover's melodious song and its breeding habitat of beaches, which we humans associate with relaxation, one might naïvely think that all is well for this species.

While habitat use by some species is best described and measured in terms of two dimensions, for other species it is useful to think linearly. Soon after they return to their breeding grounds in spring, male Piping Plovers establish territories that include a length of shoreline to be used for foraging and an area above the high-tide mark where some vegetation grows and where the birds nest. Breeding habitat is thus composed of strips of suitable shoreline, only some of which are actually used given the species' low numbers. The distance between nests of different pairs varies from about 14 to 400 m.

This species has suffered major declines as a result of competition with humans for breeding and wintering habitat. The current global population of fewer than 10,000 individuals is roughly equally split between the Atlantic Coast's beaches and the prairies, where Piping Plovers breed at lakes (alkali or freshwater), reservoirs, and rivers. Only about 1% of the population breeds at the Great Lakes. The key threats to this ground-nesting species are predation on eggs and chicks, human disturbance, and habitat loss or degradation. On the positive side of this story is the army of people and organizations that has been working on the ground for decades across the species' range to protect habitat and reduce predation. Every year, predator exclosures are placed around nests and areas of beach are cordoned off. Interpretation programs are given to educate the general public (for example regarding pet management), and habitats are restored or protected through purchase or agreements. In the prairies, where a major issue is water management, ongoing lobbying for this and other species is helping greatly. Overall, the population has been increasing in the past two or three decades as a result of these massive efforts, and it is critical that they continue because the positive population trend depends on this intensive work (Elliott-Smith and Haig 2004).

Darroch Whitaker

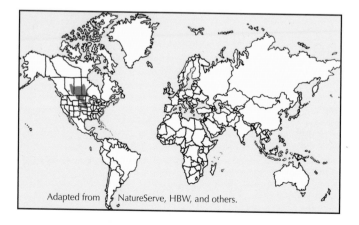

Adapted from NatureServe, HBW, and others.

NatureServe
Conservation Status

Global: .... Vulnerable

Canada: .... Vulnerable

US: .... Vulnerable

Mexico: .... Not Applicable

www.natureserve.org

Killdeer

Scientific: *Charadrius vociferus*
Français: Pluvier kildir
México: Chorlo Tildío
Order: Charadriiformes
Family: Charadriidae

The Killdeer is the world's only shorebird with two breast-bands—some individuals even show three for a brief time during molt. While most shorebirds are long-distance migrants, the Killdeer is a permanent resident throughout much of the southern part of its North American range. It is the earliest North American shorebird to return to its breeding grounds in spring. The name "Killdeer" is a phonetic imitation of its loud cry, while the species name *vociferus* means "loud".

All plovers use distraction displays to lure intruders away from the nest or chicks, but the species that most often fools humans with this age-old trick, also known as the broken wing act, is the Killdeer. "One wing was held extended over the back, the other beat wildly in the dust; the tail feathers were spread and the bird lay flat on the ground, constantly giving a wild alarm note. This performance continued until the observer came very near when the bird would rise and run along the ground in a normal manner or at most with one wing dragging slightly as long as pursuit was continued. If the observer turned back toward the nest, however, these actions were immediately repeated. When the parents had succeeded in luring the intruder about 100 yards [90 m], they seemed satisfied as they then flew away" (Bent 1929, 208). A Killdeer may also pretend to incubate eggs or brood chicks on the ground away from the nest, leading predators to believe that the nest lies under it rather than yonder (Jackson and Jackson 2000).

Edgar T. Jones

Emily Pipher

Marie-Anne Hudson

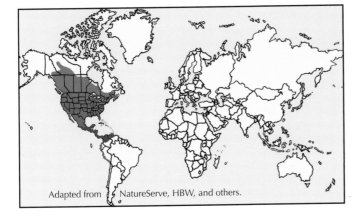

Adapted from NatureServe, HBW, and others.

NatureServe
Conservation Status

Global: .... Secure

Canada: .... Secure

US: .... Secure

Mexico: .... Secure

www.natureserve.org

Mountain Plover

Scientific: ***Charadrius montanus***
Français: Pluvier montagnard
México: Chorlo Llanero
Order: Charadriiformes
Family: Charadriidae

The Mountain Plover is not a fearful species, often facing away from an observer and squatting motionlessly when disturbed. It is easy to miss the drab "Prairie Ghost". A better name for it might have been "*Rocky* Mountain Plover" because its breeding stronghold lies in the dry plains of the Rocky Mountain plateau rather than in the mountains. Here the vegetation is kept low and sparsely distributed by natural disturbances such as fire and herbivory (e.g., from Black-tailed Prairie Dogs (*Cynomys ludovicianus*), American Bison (*Bos bison*), and Pronghorns (*Antilocapra americana*)) or by agricultural machinery. Controlled burning of habitat has resulted in some success in bringing back breeding or wintering Mountain Plovers. In some cases birds have even reappeared on fields that were still smoldering. In the long term, conservation needs for this species include reversing efforts to reduce prairie dog populations and continuing the work being done to bring back American Bison via reintroduction.

In the Mountain Plover's mating system, the female typically lays six eggs, three in one nest and three in another. The male incubates the first nest and the female incubates the second one, the two birds keep a distance of 6–15 m between them, and they don't defend a territory together. However, in one study three nests each contained a female's full complement of six eggs and all 18 eggs hatched—thus some behavioral plasticity exists. In the two-nest situation, nests tended by males have higher hatching success than those tended by females. After

Billl Schmoker

hatching, the tendency is for only one parent to care for the brood at least until they can fly, and if the female is the caretaker she will likely have a higher rate of success in rearing them to fledging than will the male. When biparental care occurs, the brood is usually split between the parents. Continuing with the theme of disturbance in relation to habitat and behavior above, we can think of female and male Mountain Plovers not wanting to "disturb" each other, so they separate breeding tasks and then find other mating partners next year (Knopf and Wunder 2006).

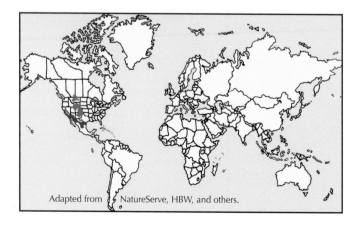

Adapted from NatureServe, HBW, and others.

NatureServe
Conservation Status

Global: .... Vulnerable
Canada: .... Critically Imperiled
US: .... Vulnerable
Mexico: .... Imperiled

www.natureserve.org

Eurasian Dotterel

Scientific: *Charadrius morinellus*
Français: Pluvier guignard
México: Chorlo Carambolo
Order: Charadriiformes
Family: Charadriidae

"The regularity of the appearance of the Eurasian Dotterel on migration at almost the same spots [in Europe] year after year and at approximately the same time in spring has proved a great disadvantage to the species. Being very tame and unsuspicious and much sought after, not only as a delicacy for the table, but also on account of the demand for its feathers on the part of fly-fishers, it was mercilessly shot on the way to its breeding grounds in the north of England and Scotland" (Jourdain, F. C. R., in Bent 1929, 150). This species' unwary nature has also led to the term "dotterel" being directed at some humans and meaning "a doting, old fool".

Most plover species are seasonally monogamous, meaning that they generally stay with the same mate during a breeding season, though often not from one season to the next. However, some are sequentially polyandrous, meaning that the female will mate first with one male, then with another during the same breeding season. As we saw in the previous account, female Mountain Plovers often abandon the male soon after laying eggs for him to incubate. Then they produce three more eggs in another nest, which they incubate themselves. The female Eurasian Dotterel goes one step further by deserting the first male, as does the Mountain Plover, but then mating with a second male whom she also abandons after laying eggs for him to incubate.

Vaughan Ashby

Thus female dotterels rarely help raise chicks. However, when population sex ratios are skewed, or in other years for unknown reasons, Eurasian Dotterels are polygynous, meaning that one male will breed with more than one female (Sibley 2001).

The Eurasian Dotterel occurs in northwestern Alaska in arctic tundra and arctic-alpine zones where it prefers unvegetated, flat areas. Only small numbers are thought to breed there and its state rank is Apparently Secure. Because of the remoteness of its breeding range, it "is undoubtedly under-surveyed in Alaska" (Tracey Gotthardt, Alaska Natural Heritage Program, pers. comm.) The collection of 10 specimens before 1959 suggests that it may have nested regularly in Alaska for some time.

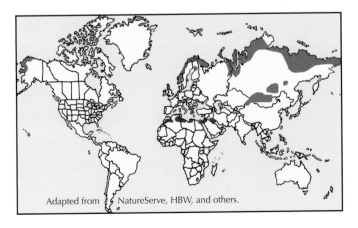

Adapted from NatureServe, HBW, and others.

NatureServe
Conservation Status

Global: .... Secure

Canada: .... Not Applicable

US: .... Vulnerable

Mexico: .... Not Applicable

www.natureserve.org

Northern Jacana

Scientific: *Jacana spinosa*
Français: Jacana du Mexique
México: Jacana Norteña
Order: Charadriiformes
Family: Jacanidae

Jacanas have amazingly long toes and toenails that allow them to walk over floating vegetation in freshwater marshes. When they look like they are walking on water, they are probably treading on submerged vegetation. Northern Jacanas have a sharp, horny, spur at the bend of each wing that they use in fighting. Hence the species name *spinosa*, meaning "spine".

Jacanas exhibit a very high degree of reversed sexual size dimorphism. In breeding adult Northern Jacanas, for example, females average 161 g compared to 91 g in males. Males build the nest, incubate the eggs, and rear the young. Females are more aggressive than males and may have up to four males nesting for them at the same time. While the males under one female defend their adjacent territories against one another, the females defend their larger territories against other females and also help each of their males defend their territories against intruders of the same or other species. Competition for males and territories can be so intense that a female that usurps a territory and the males from another female may destroy the latter's nests and chicks, presumably to increase the probability that the males will mate with her right away. This simultaneous polyandry occurs when the habitat has considerable food, which allows male and female territories to be small. In less rich habitats females may be able to support only one male, whose territory will likely be large (Jenni and Mace 1991).

bryanjsmith

The Northern Jacana, a fairly widespread species in Mexico, is a very rare visitor to the Lower Rio Grande Valley of Texas where it is seen mostly in winter. Of the 36 records accepted by the Texas Bird Records Committee of the Texas Ornithological Society, which span the period 1889–2012, only six are from 2006 or later. However, a population of more than 40 birds at Maner Lake, Brazoria County, south of Houston, existed from 1967 to 1978. Breeding was apparently occurring regularly, thus justifying the inclusion of this species in the main part of this book (Lockwood and Freeman 2014).

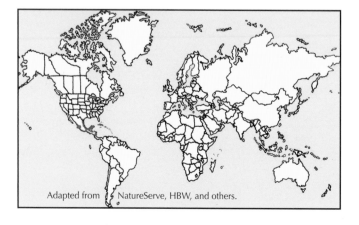

Adapted from NatureServe, HBW, and others.

NatureServe
Conservation Status

Global: .... Secure
Canada: .... Not Applicable
US: .... Not Applicable
Mexico: .... Secure

www.natureserve.org

Spotted Sandpiper

Scientific: *Actitis macularius*
Français: Chevalier grivelé
México: Playero Alzacolita
Order: Charadriiformes
Family: Scolopacidae

While paddling hundreds of kilometers of lake shorelines searching for loon nests, I flushed single Spotted Sandpipers on countless occasions, yet cannot improve on the following description. "The young Spotted Sandpiper furnishes an instance of an ancestral habit springing into action almost at the moment of hatching. When no larger than the egg from which they have just stepped, they run over the sand teetering their tail in the manner of their parents.... Almost every inhabitant of the United States sometime during the year may meet this graceful little wader stepping delicately along the margin of some sandy pond, the shore of the sea, or skimming from perch to perch on the rocks bordering a mountain stream.... Except when creeping up within reach of an insect or when its attention is riveted on the snapping up of a bit of food, the tail is almost continuously in motion up and down. At the least alarm, the motion is increased to a wider arc until the posterior half of the bird's body is rapidly teetering. A little increase in alarm and the bird is off on vibrating wings held stiffly and cupped with the tips depressed, sailing along the shore away from danger.... When it first starts from the shore, the wings seem to vibrate like a taut wire; then, as the bird gains headway, they set and, depressed and quivering, they carry the bird slowly onward, often swaying from side to side close to the surface of the water" (Tyler, W. M., in Bent 1929, 79–87). As a shorebird exhibiting reversed sex roles, the Spotted Sandpiper was the first migratory bird species in which females were shown to arrive on the breeding grounds in spring before males. It was also the first bird species in which males were found to have higher levels of prolactin, a hormone that promotes parental care, than females (Reed et al. 2013).

Dennis Paulson

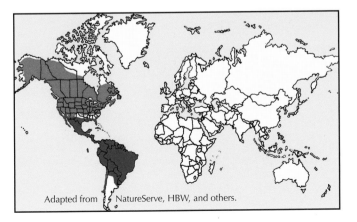

Adapted from NatureServe, HBW, and others.

NatureServe
Conservation Status

Global:....................... Secure

Canada:...................... Secure

US:............................. Secure

Mexico:......... Not Applicable

www.natureserve.org

Solitary Sandpiper

Scientific: *Tringa solitaria*
Français: Chevalier solitaire
México: Playero Solitario
Order: Charadriiformes
Family: Scolopacidae

"Solitary Sandpiper" is a good name for this species that migrates in small groups at best, a habit that saved it from being shot during the market-hunting days when the idea was to kill as many birds as possible with one discharge of shot. It breeds near ponds and pools in areas of muskeg bogs and coniferous trees in the boreal forest. The following discussion is an excellent example of the sort of error that can become perpetuated in the scientific or naturalist literature. "The nesting habits of this sandpiper long remained a mystery or were misunderstood.... I came across no less than seven published records of nests found on the ground and said to be positively identified as this species. These were all published prior to the discovery of the now well-known habit of nesting in the deserted nests of passerine birds. Not a single one of these records seems to be substantiated by an available specimen of the parent bird" (Bent 1929, 2). This is the only North American shorebird that regularly nests in trees. It is considered to be the sister species of the Green Sandpiper (*Tringa ochropus*) of the Old World, which does the same.

In the current age of haste, rarely do we read such thoughtfully written descriptions based on hours of close observation as the following. "It advances one foot at a time and by rapidly

Edgar T. Jones

moving the forward foot, stirs up the vegetation at the bottom ever so slightly. This motion … is done with intent to disturb insects among the algae at the bottom without roiling the water, and the eager bird leaning forward plunges in its bill and head sometimes to the eyes and catches the alarmed water insects as they dart away. I have watched this carefully with a glass [probably binoculars] while lying in the grass only 10 or 12 feet [3–4 m] from the bird. It is easy by stirring the bottom slightly with a stick to cause a similar movement of the water insects, but I never could agitate it so delicately as to avoid clouding the water with sediment from the bottom" (Forbush 1912, 308).

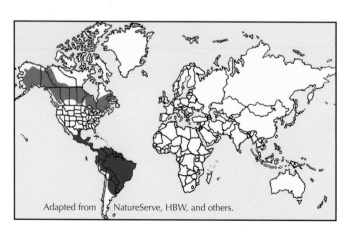

Adapted from NatureServe, HBW, and others.

NatureServe
Conservation Status

Global: .... Secure
Canada: .... Secure
US: .... Apparently Secure
Mexico: .... Not Applicable

www.natureserve.org

Wandering Tattler

Scientific: *Tringa incana*
Français: Chevalier errant
México: Playero Vagabundo
Order: Charadriiformes
Family: Scolopacidae

With a global population that may number as few as 10,000 individuals, the Wandering Tattler is one of the least common North American shorebirds. Usually only one to three individuals are seen at a time, but flocks of up to several dozen occur when the birds are roosting or when they have just landed during migration. Thus it is not a very well-known species. Its dull gray upperparts combined with the similar coloration of much of its habitat do not help in its detection. Very few people see it in its northern breeding range. For most Americans and Canadians, it is most easily observed on the west coast outside of the breeding season.

In North America, the Wandering Tattler breeds on the alpine tundra, mostly in Alaska and the Yukon Territory, where it forages and roosts along streams and lakes. The ground may be rocky or covered with moss and/or dwarf shrubs. Damp meadows are also used. The nest is usually located on a rocky or gravelly site, often near water. When I participated for a day in an early-spring bird count in the Yukon near Whitehorse and reported during the evening tally that I thought I had seen my first-ever Wandering Tattler, the question the local pros asked me was, "Did you see it foraging at the edge of lake ice?" When I answered "Yes", they were reassured.

During migration and in winter, however, this species is found in the rocky intertidal zone in natural situations and at piers, jetties, and rock pilings. Because much of the nonbreeding range is in the South Pacific Ocean, this huge area's atolls and islands should be studied

bryanjsmith

Sara Acosta

as possible havens for Wandering Tattlers and also for rare, isolated, sessile biota. Habitat restoration combined with solid legal protection, which in many such areas is now in place, could lead to a critical network of conservation areas (Gill et al. 2002).

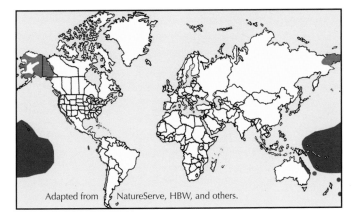

Adapted from NatureServe, HBW, and others.

NatureServe
Conservation Status

Global: .... Secure
Canada: .... Vulnerable
US: .... Apparently Secure
Mexico: .... Not Applicable

www.natureserve.org

Greater Yellowlegs

Scientific: *Tringa melanoleuca*
Français: Grand Chevalier
México: Patamarilla Mayor
Order: Charadriiformes
Family: Scolopacidae

"The names 'telltale' and 'tattler' have long been applied to both of the yellowlegs [Greater and Lesser] and deservedly so, for their noisy, talkative habits are their best known traits. They are always on the alert and ever vigilant to warn their less observant or more trusting companions by their loud, insistent cries of alarm that some danger is approaching. Every sportsman [hunter] knows this trait and tries to avoid arousing this alarm when other more desirable game is likely to be frightened away. And many a yellowlegs has been shot by an angry gunner as a reward for his exasperating loquacity" (Bent 1927, 321).

In spring, a portion of the Greater Yellowlegs population remains on the wintering grounds rather than migrating back to their muskeg breeding habitat. Why? A study in coastal Venezuela found that many of the seasonal wetlands used by this species in the nonbreeding season became drier and smaller from January to April because of reduced rainfall at that time of year. The birds became crowded, thus giving internal parasites such as flatworms an excellent opportunity to infest more birds. Juvenile birds that had hatched in the previous year up north had arrived in their wintering grounds largely free of these parasites, whereas many adults had arrived from their southward migration infested. However, by spring migration-time, juveniles were more heavily infested than adults, which may have developed an immunity after having

Bill Schmoker

survived previous infestation. The only birds that could migrate back north seemed to be ones whose parasite load was low enough to allow them to put on enough fat and flight feathers to migrate. Thus, only parts of this puzzle are firmly established, and further study should test these hypotheses to determine whether the entire story holds true. If so, the problem is being exacerbated in this, and perhaps other species, by reductions in such shorebird wintering habitats occurring worldwide (McNeil et al. 1995).

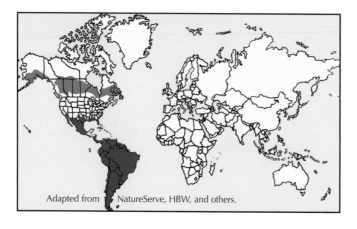

Adapted from NatureServe, HBW, and others.

NatureServe
Conservation Status

Global: .... Secure

Canada: .... Secure

US: .... Secure

Mexico: .... Not Applicable

www.natureserve.org

Willet

Scientific: *Tringa semipalmata*
Français: *Chevalier semipalmé*
México: Playero Pihuiuí
Order: Charadriiformes
Family: Scolopacidae

The dull-colored Willet first presents itself as a difficult bird to identify given its lack of striking features—that is until it opens its wings and displays a broad white wing-stripe bordered by black, a feature visible at any time of year. The identification is clinched by the bird's English cry of *pill-will-willet*, in which its name is cleverly imbedded. But it is a bilingual species, and when it exits the US southward over the Mexican customs office, it changes its call to *pihuiuí* for the benefit of our Spanish-speaking neighbors.

The Willet's global breeding distribution consists of two widely separated populations in North America: the Western Willet (WW) of the Prairies and the Eastern Willet (EW) of the East Coast. Population estimates from winter counts are 160,000 WWs and 90,000 EWs. WWs often have abundant choices of places to nest, but food availability is unpredictable because habitat features vary greatly with rainfall patterns there. Conversely, EWs often have abundant food combined with fewer places to nest. Freshwater prairie marshes are typically less productive than coastal saltwater marshes, which are constantly replenished by tidal movements. Thus, WWs tend to have larger foraging territories than do EWs (Lowther et al. 2001).

Dennis Paulson

Juan Bahamon

An atypical EW colony breeds in a sphagnum bog in Corea, Maine, which is described as having hummocks covered by abundant lichen and sedges. Normally the EW nests in or near salt marshes or in sand dune areas (Wells and Vickery 1990).

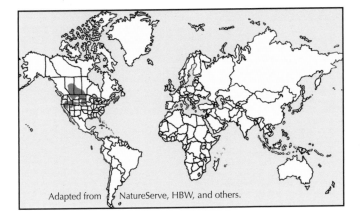

Adapted from NatureServe, HBW, and others.

NatureServe
Conservation Status

Global: .... Secure

Canada: .... Secure

US: .... Secure

Mexico: .... Secure

www.natureserve.org

Lesser Yellowlegs

Scientific: *Tringa flavipes*
Français: Petit Chevalier
México: Patamarilla Menor
Order: Charadriiformes
Family: Scolopacidae

When you see a "one-legged" shorebird and are wondering where the other leg is, it is probably tucked up out of view under some feathers to conserve heat. Considerable energy is lost from unfeathered legs, and this can be greatly reduced by keeping one of them out of contact with the ground and the wind. "Like many other shorebirds, they are fond of standing on one leg or even hopping about on it for a long time as if one leg were missing. Often a number of birds will be seen all doing this at the same time as if playing a sort of game, but if we watch them long enough the other leg will come down, for they are not cripples" (Bent 1927, 273).

An estimated 400,000 Lesser Yellowlegs exist. Various breeding, wintering, and migration studies suggest population declines over the past few decades. Generally speaking, these efforts have not been coordinated with each other, so various methodologies have been used, though mostly with the same result. Nevertheless, the decline has not been considered serious enough by NatureServe or the International Union for Conservation of Nature and Natural Resources (IUCN) to assign this species to any "at risk" category. There is usually some concern for wild species, however, regardless of their rankings. Thus, Birdlife International, the world's largest conservation network, has identified 18 sites as qualifying for the highest priority for this species in their system of Important Bird Areas (IBAs), that is sites of global importance—they hold at least one percent of the global population at some time during the year. Of note, the two high-priority Barbados IBAs on that list have the term "shooting swamp" in their name. Illegal hunting continues there and on other Caribbean islands, with 7,000–15,000 birds taken annually during autumn migration in the Barbados alone. This is a perceived threat to the global population (Clay et al. 2012).

Dennis Paulson

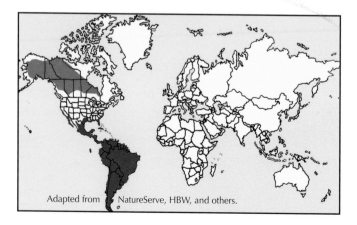

Adapted from NatureServe, HBW, and others.

NatureServe
Conservation Status

Global: .... Secure

Canada: .... Secure

US: .... Secure

Mexico: .... Not Applicable

www.natureserve.org

Upland Sandpiper

Scientific: *Bartramia longicauda*
Français: Maubèche des champs
México: Zarapito Ganga
Order: Charadriiformes
Family: Scolopacidae

"It is a characteristic bird of the prairies and wide open grassy fields, where it once abounded in enormous numbers" (Bent 1929, 55–56). "About 1880, when the supply of Passenger Pigeons began to fail, and the marketmen, looking about for some other game for the table of the epicure in spring and summer, called for plover, the destruction of the "Upland Plover" [now called Upland Sandpiper] began in earnest. The price increased. In the spring migration the birds were met by a horde of market gunners, shot, packed in barrels, and shipped to the cities. There are tales of special refrigerator cars sent out to the prairie regions, and parties of gunners regularly employed to follow the birds and ship plover and curlews by the carload to the Chicago market. These may not be based on facts, but we know that the birds came to market in great quantities" (Forbush 1912, 317). Fortunately, we didn't lose this species. Its numbers have been dropping and habitat loss in some parts of its range has compelled it to try breeding in non-traditional sites such as commercial blueberry barrens, bogs, and airports, where its reproductive success is unknown (Houston et al. 2011).

"Practically 97% of the food [of the Upland Sandpiper] consists of animal matter, chiefly of injurious and neutral forms. The vegetable food comprises the seeds of such weed pests as buttonweed, foxtail grass, and sand spurs, and hence

Edgar T. Jones

is also to the credit of the bird" (McAtee 1912, 16). A recurring theme in Bent's encyclopedic *Life Histories of North American Birds* series was the "value" of each bird to humans: was it an asset, a menace, or neither? He took the same approach with the birds' food species, be they animal or plant. Some critics might view this approach as simplistic, but much of the information generated served as a starting point for many later studies. I think Mr. Bent knew, perhaps better than anyone today, the intrinsic and ecological value of each species, including the Upland Sandpiper.

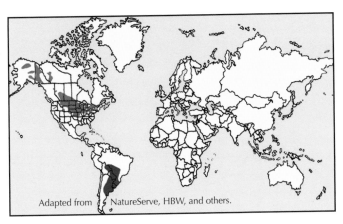

Adapted from NatureServe, HBW, and others.

NatureServe
Conservation Status

Global: .... Secure

Canada: .... Secure

US: .... Secure

Mexico: .... Not Applicable

www.natureserve.org

Eskimo Curlew

Scientific: ***Numenius borealis***
Français: Courlis esquimau
México: Zarapito Boreal
Order: Charadriiformes
Family: Scolopacidae

"The aerial evolutions of the curlews when migrating are perhaps one of the most wonderful in the flight of birds.... These flocks reminded the settlers of the flights of Passenger Pigeons, and the curlews were given the name of 'prairie pigeons'.... The gunner's name for the Eskimo Curlew was 'dough-bird', not 'doe-bird', for it was so fat when it reached us in the fall that its breast would often burst open when it fell to the ground, and the thick layer of fat was so soft that it felt like a ball of dough. [In the late 1800s] the slaughter of these poor birds was appalling and almost unbelievable. Hunters would drive out from Omaha and shoot the birds without mercy until they had literally slaughtered a wagonload of them, the wagons being actually filled, and often with the sideboards on at that. Sometimes when the flight was unusually heavy and the hunters were well supplied with ammunition, their wagons were too quickly and easily filled, so whole loads of the birds would be dumped on the prairie, their bodies forming piles as large as a couple of tons of coal, where they would be allowed to rot while the hunters proceeded to refill their wagons with fresh victims and thus further gratify their lust of killing.... So dense were the flocks when the birds were turning in their flight that one could scarcely throw a brick or missile into it without striking a bird" (Turner, L. M., Swenk, M. H., Bent, A. C., in Bent 1929, 128–134).

Other factors in this species' demise were extensive grassland loss and the extinction of a major food source in spring migration: Rocky Mountain Locust (*Melanoplus spretus*) eggs, larvae, and adults. Searches for Eskimo Curlews continue (Wells 2007). If one is found, the lucky person who can prove it may become as much of a celebrity as those who break North American "Big Year" records.

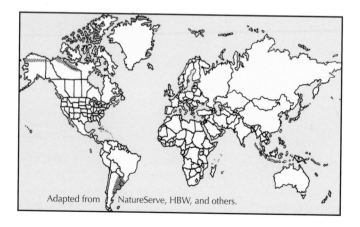

Donald Bleitz (Courtesy of the Western Foundation of Vertebrate Zoology)

One of the last photos taken of this species in the wild: Galveston Texas, 1962

Adapted from NatureServe, HBW, and others.

NatureServe
Conservation Status

Global: ....Possibly Extinct
Canada: .....Possibly Extirpated
US: ....Possibly Extirpated
Mexico: .....Possibly Extirpated

www.natureserve.org

Whimbrel

Scientific: *Numenius phaeopus*
Français: Courlis corlieu
México: Zarapito Trinador
Order: Charadriiformes
Family: Scolopacidae

Of the world's eight species of "curlews" that make up the genus *Numenius*, only the Whimbrel is missing the word "curlew" in its English name. On the other hand, the stone-curlews (family Burhinidae) are not part of this group (though they are shorebirds). "Curlews" are mottled brown and have long, downcurved bills. In the case of Whimbrels, the overall diet is broad, including fish, insects, and berries. However, the main foods are marine invertebrates, and the bill is adapted to foraging for them in the intertidal zone. When hunting crabs, which flee upon the slightest Whimbrel movement to scurry into burrows, a Whimbrel will usually probe the entrance and rotate the bill if necessary to follow the burrow's curve in an attempt to secure the crab. Meanwhile the crab is likely trying to wedge itself into the end of the burrow to avoid being eaten. Larger crabs probably wedge themselves in more strongly than do small ones. In a migration study, bill shape nicely matched the shape of crab burrows in Panama; the match was almost as good in Cape Cod. Whimbrels obtain food by inserting the bill to varying degrees in substrate, sometimes just the tip and other times more than half the bill, or anything between (Skeel and Mallory 1996).

Robert Alvo

Whimbrel numbers seem to have been increasing in Europe. In central Finland, for example, they have begun breeding on farmlands (Grant and Väisänen 1997). On the Shetland Islands of Scotland they breed in heathlands, many of which are lightly grazed by sheep, cattle, and ponies. Re-seeding with grass mixtures by humans to improve grazing by large mammals is acceptable for nesting Whimbrels as long as the existing vegetation is not removed by ploughing or harrowing (Grant 1992).

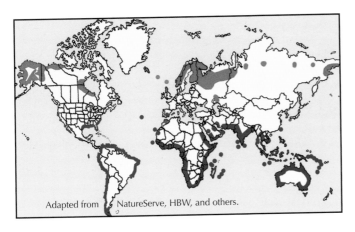

Adapted from NatureServe, HBW, and others.

NatureServe
Conservation Status

Global: .... Secure
Canada: ..... Apparently Secure
US: .... Secure
Mexico: .... Not Applicable

www.natureserve.org

Bristle-thighed Curlew

Scientific: *Numenius tahitiensis*
Français: Courlis d'Alaska
México: Zarapito del Pacífico
Order: Charadriiformes
Family: Scolopacidae

The Bristle-thighed Curlew's English name is the only one of the four names listed above that refers to this species' long, barbless, thigh-feather-shafts of unknown function. This is the only shorebird that cannot fly while it is molting. It is also the only migratory shorebird that winters exclusively on oceanic islands.

Another of this curlew's unique features among the shorebirds is its skill with tools—using its bill, it throws fragments of coral at eggs to break them. "That a bird of the shorebird family should destroy eggs may seem almost unbelievable in view of the habits ordinary in this group, yet in work in the Hawaiian Bird Reservation in 1923 we found the Bristle-thighed Curlew and the Ruddy Turnstone making regular practice of eating the eggs of the birds nesting on these distant islands… A curlew deliberately opened an old albatross egg found in the sand and ate eagerly from the putrid interior. As this egg had been lying unprotected from the sun for at least four months previous, its condition may be imagined" (Wetmore, A., in Bent 1929, 142–143). Less nauseating but more disturbing, a curlew on Laysan Island tried several times to swallow a 2-cm-diameter, white, plastic ball, probably thinking it was an egg. Egg predation by this species may be opportunistic and uncommon because the few accounts of it involve situations in which insect prey availability may have been low as a result of greatly reduced vegetation

Dennis Paulson

due to foraging by introduced European Rabbits (*Oryctolagus cuniculus*) (Marks et al. 2002).

The Bristle-thighed Curlew's Imperiled global status assessment is based on the species' very limited breeding range (two small areas in western Alaska), its estimated population of 10,000 birds, (including only an estimated 3200 breeding pairs), its requirement of 3–4 years before young birds first migrate north to Alaska, and possible threats on its wintering islands (NatureServe 2015).

Adapted from NatureServe, HBW, and others.

NatureServe
Conservation Status

Global: .... Imperiled

Canada: .... Not Applicable

US: .... Imperiled

Mexico: .... Not Applicable

www.natureserve.org

ROBERT ALVO

Long-billed Curlew

Scientific: *Numenius americanus*
Français: Courlis à long bec
México: Zarapito Pico Largo
Order: Charadriiformes
Family: Scolopacidae

This is the last of our four "curlews", whose genus name *Numenius* means "new moon", referring to the crescent-shaped moon of which the downward-curved bill is reminiscent. The largest of our shorebirds, it "seems to embody more than any other the wild, roving spirit of the vast, open prairies. Its large size, its long, curving bill, the flash of cinnamon in its wings, and above all, its loud, clear, and prolonged whistling notes are bound to attract attention" (Bent 1929, 97–98). Symbolically, the Long-billed Curlew is drawing attention to the grassland's plight as the most endangered ecosystem in North America, and to the efforts of numerous partners to restore the natural processes with which humans have interfered but that keep grasslands from becoming forests: fire, drought, and moderate wildlife-grazing (see Mountain Plover account, p. 178). Apart from the curlew, the other eight birds that are endemic breeders of the Great Plains, all of which have suffered significant declines, are the Ferruginous Hawk (the only tree-nester) and seven ground-nesters (the Mountain Plover and six passerines). Most of our grasslands are gone, so protection and proper management of the remaining bits are critical.

Juan Bahamon

The carnivorous Long-billed Curlew eats mainly invertebrates but also opportunistically some vertebrates such as bird eggs and nestlings in its breeding habitat. There it does not seem to rely on the curvature of its long bill for capturing prey, which it usually picks with its bill. Rather, the long curvature seems more an adaptation to feeding on crabs and shrimps living in deep burrows on its wintering grounds, where it probes for them (Dugger and Dugger 2002) (see Whimbrel account, p. 189).

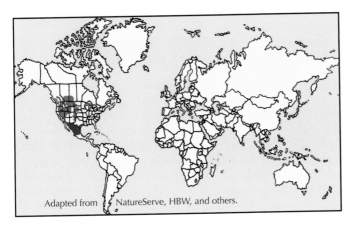

Adapted from NatureServe, HBW, and others.

NatureServe
Conservation Status

Global: .... Secure
Canada: .... Vulnerable
US: .... Secure
Mexico: .... Not Applicable

www.natureserve.org

Hudsonian Godwit

Scientific: *Limosa haemastica*
Français: Barge hudsonienne
México: Picopando del Este
Order: Charadriiformes
Family: Scolopacidae

Like our two "tattletales", the Greater Yellowlegs and Lesser Yellowlegs, the Hudsonian Godwit likes to stand on treetops on its breeding grounds, calling loudly with its wings raised while constantly readjusting its position to stay balanced, for it cannot curl its toes around the branches. Parents also call repeatedly when their agitated chicks are making distress calls, apparently to muffle or distort the chicks' sounds in an effort to make them more difficult for predators to locate.

This species is known to breed in three widely separated regions: southwestern Alaska, the northern Northwest Territories, and western Hudson Bay. However, it must breed elsewhere because the numbers of birds estimated for these known regions do not nearly add up to the estimates obtained outside the breeding season. On the one hand, many potential areas between these regions and outside of them have not been surveyed, while on the other hand many other such areas that have been surveyed have not yielded any godwits. New breeding areas are still being found. Perhaps some key aspect of its breeding habitat remains unknown. Or—might it breed in southern South America, unbeknownst to biologists? It occurs there throughout the year....

None of the Hudsonian Godwit's characteristics (e.g., plumage coloration, morphometrics) are known to vary across the species' breeding range, and thus no subspecies have been described. However, genetic studies have shown marked differences between birds of the three breeding regions, suggesting that further investigation could reveal differences in characteristics that have not been adequately compared, such as vocalizations. Conservation of this species could be enhanced considerably by actions based on a better knowledge of its breeding distribution and the extent to which genes flow among the various populations (Walker et al. 2011).

Dennis Paulson

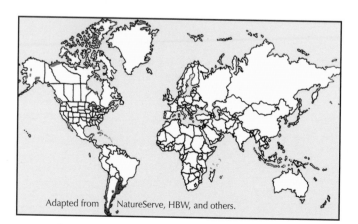

Adapted from NatureServe, HBW, and others.

NatureServe
Conservation Status

Global: ...... Apparently Secure

Canada: ..... Apparently Secure

US: .... Vulnerable

Mexico: .... Not Applicable

www.natureserve.org

Bar-tailed Godwit

Scientific: *Limosa lapponica*
Français: Barge rousse
México: Picopando Cola Barrada
Order: Charadriiformes
Family: Scolopacidae

If you needed to migrate from southwestern Alaska to New Zealand or eastern Australia every year, you might consider the vast Pacific Ocean as a formidable barrier. Yet recent studies suggest that much of the Bar-tailed Godwit population breeding in Alaska flies nonstop for five to nine days and up to 11,000 km, thus winning the nonstop flight record of the world's birds. The Pacific Ocean in fact seems to be an excellent migration corridor for this species. At the end of the breeding season the birds capitalize on the rich feeding conditions at the Yukon–Kuskokwim Delta in southwestern Alaska to put on fat such that they have the highest proportion of body mass as fat (55%) known in migrating birds. By making the long flight nonstop, the godwits avoid predators, parasites, and diseases on the way. The predictable weather systems provide the birds with favorable winds during the first part of the voyage, and islands in the South Pacific offer them landing sites if required later on. One concern for this species with respect to global climate change is more variable weather conditions, particularly during migration, because we do not know how flexible these birds are to such changes (Gill et al. 2008).

"The Bar-tailed Godwit differed from the other shorebirds nesting at Hooper Bay, Alaska, in that individuals in immature plumage were breeding. Sometimes a gray-breasted immature female would be paired with a rich plumaged male, or again both mates would be in full color;

Dennis Paulson

but I encountered many pairs in which both parents showed the light grayish breast of adolescence" (Brandt, H. W., in Bent 1927, 294). Researcher Brian J. McCaffery wrote me that other early ornithologists noted the same phenomenon, but that what they considered "immature" plumage was simply within the range of typical breeding plumages on the Yukon-Kuskokwim Delta. These breeders are never as richly plumaged as many males breeding in northern Alaska. The brighter birds are seen as they migrate through southern Alaska after the local breeders are established on territories.

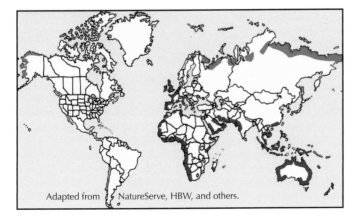

Adapted from NatureServe, HBW, and others.

NatureServe
Conservation Status

Global: .... Secure
Canada: .... Not Applicable
US: .... Secure
Mexico: .... Not Applicable

www.natureserve.org

Marbled Godwit

Scientific: *Limosa fedoa*
Français: Barge marbrée
México: Picopando Canelo
Order: Charadriiformes
Family: Scolopacidae

Unlike our other two godwits, the Marbled Godwit is mainly a prairie-breeding species, although it has two small, disjunct northern populations—in southern Alaska and at James Bay. During autumn staging and on the wintering grounds, it flocks with other large shorebirds such as Long-billed Curlews, Hudsonian Godwits, Whimbrels, and Willets. Although uncharacteristic for shorebirds, the Marbled Godwit at times during migration feeds practically exclusively on plant tubers, as does the Hudsonian Godwit.

For some decades, many shorebird species have been suffering major population declines (North American Bird Conservation Initiative 2012). Most shorebird species, including the Marbled Godwit, gather in huge numbers at relatively few sites during migration and in winter. In 1982, Guy Morrison of the Canadian Wildlife Service saw this as an opportunity, suggesting that a "series of protected areas linking key sites" be created for the Western Hemisphere, based in part on an atlas of South American wintering sites of shorebirds breeding in Canada, which he and colleague Ken Ross had completed. Indeed, the year 2015 marks the 30th anniversary of the formalization of The Western Hemisphere Shorebird Reserve Network (WHSRN), an international strategy to protect shorebirds and their most important sites. WHSRN designated its first site, Delaware Bay in the US states of New Jersey and Delaware, in 1986. Today 90 sites have been designated in 13 countries such that more than 13 million hectares of shorebird habitat are recognized (http://www.whsrn.org). The WHSRN site with the largest estimate of Marbled Godwits (69,000 birds, representing 35–50% of the global population) is Ojo de Liebre/ Guerrero Negro in central Baja California, Mexico, in winter (Melcher et al. 2010).

Dennis Paulson

Emily Pipher

Most shorebirds nest on the ground.

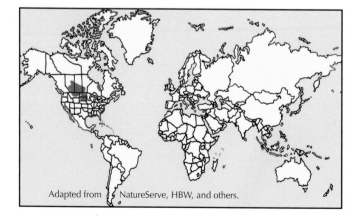
Adapted from NatureServe, HBW, and others.

NatureServe
Conservation Status

Global: .... Secure

Canada: .... Secure

US: .... Secure

Mexico: .... Not Applicable

www.natureserve.org

Ruddy Turnstone

Scientific: ***Arenaria interpres***
Français: Tournepierre à collier
México: Vuelvepiedras Rojizo
Order: Charadriiformes
Family: Scolopacidae

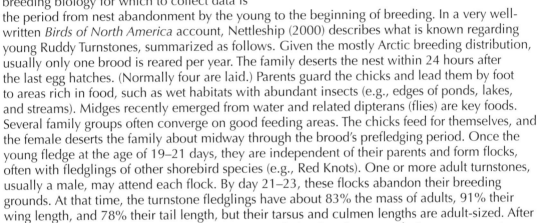

The Ruddy Turnstone's species name *interpres* means an interpreter or go-between, referring to this shorebird's varied vocalizations that warn other birds of danger (Donovan and Ouellet 1993).

One of the more difficult aspects of bird breeding biology for which to collect data is the period from nest abandonment by the young to the beginning of breeding. In a very well-written *Birds of North America* account, Nettleship (2000) describes what is known regarding young Ruddy Turnstones, summarized as follows. Given the mostly Arctic breeding distribution, usually only one brood is reared per year. The family deserts the nest within 24 hours after the last egg hatches. (Normally four are laid.) Parents guard the chicks and lead them by foot to areas rich in food, such as wet habitats with abundant insects (e.g., edges of ponds, lakes, and streams). Midges recently emerged from water and related dipterans (flies) are key foods. Several family groups often converge on good feeding areas. The chicks feed for themselves, and the female deserts the family about midway through the brood's prefledging period. Once the young fledge at the age of 19–21 days, they are independent of their parents and form flocks, often with fledglings of other shorebird species (e.g., Red Knots). One or more adult turnstones, usually a male, may attend each flock. By day 21–23, these flocks abandon their breeding grounds. At that time, the turnstone fledglings have about 83% the mass of adults, 91% their wing length, and 78% their tail length, but their tarsus and culmen lengths are adult-sized. After

David Laliberte

migrating to their wintering grounds, immature turnstones spend their first full summer there, with some moving a little northward. One-year-old birds often show partial or even complete breeding plumage, but their reproductive organs are not active. Most Ruddy Turnstones start breeding when they are two years old, although some wait until their third or fourth year. As in many other species, some young birds likely return to the breeding grounds, where some females produce few or no eggs and where some males occupy territories without breeding.

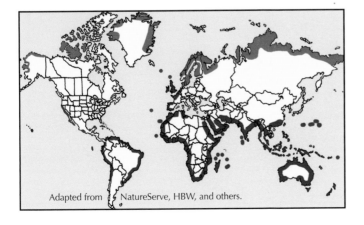

Adapted from NatureServe, HBW, and others.

NatureServe
Conservation Status

Global: .... Secure
Canada: ..... Apparently Secure
US: .... Secure
Mexico: .... Not Applicable

www.natureserve.org

Black Turnstone

Scientific: *Arenaria melanocephala*
Français: Tournepierre noir
México: Vuelvepiedras Negro
Order: Charadriiformes
Family: Scolopacidae

Only two turnstone species exist and they form a single genus: *Arenaria*. The shared part of their English name gives only a taste of the objects that they turn over to uncover their animal prey, which are mostly invertebrates.

The Black Turnstone spends most of its time outside the breeding season on seacoasts, where it forages in the wave-splash zone of shores, reefs, headlands, and sea stacks (and other coastal habitats). There it flips objects such as driftwood, sticks, shells, mud, and presumably plastic bottles and other rubbish, to access what lies underneath or on the underside. At low tide, this turnstone is generally heard before it is seen, for it is usually well camouflaged and it calls frequently. "As it stands motionless it is almost invisible in its coat of dark brown and might easily be mistaken for a knob of rock or a bunch of seaweed; but when startled into flight its conspicuous pattern of black and white flashes out a distinctive mark of recognition" (Bent 1929). When foraging on sediment such as sand on beaches, it may use its bill and forehead to flip over an edge of seaweed mat and roll it up into a cylinder. It then feeds on invertebrates that it finds on the uncovered substrate or on the underside of the seaweed.

Dennis Paulson

Throughout the year, the Black Turnstone usually occurs within two kilometers of a coast. On its preferred breeding habitat of wet sedge meadows, it finds only vegetation and driftwood to flip, and thus is often called "turnstick". Other less-preferred nesting habitats found inland include willow/sedge/reed-grass shrubland, and dwarf shrubland. This species sometimes nests up to 100 km inland along rivers or next to lakes. Like its congener, the Ruddy Turnstone, "it appears to be the self-appointed sentinel of its nesting community" (Handel and Gill 2001).

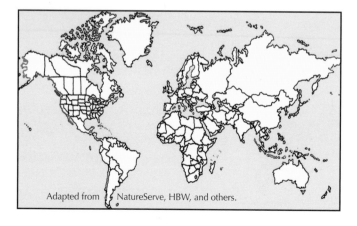

Adapted from NatureServe, HBW, and others.

NatureServe
Conservation Status

Global: .... Secure

Canada: .... Secure

US: .... Secure

Mexico: .... Not Applicable

www.natureserve.org

Red Knot

Scientific: *Calidris canutus*
Français: Bécasseau maubèche
México: Playero Rojo
Order: Charadriiformes
Family: Scolopacidae

Horseshoe crabs are not really crabs, which are crustaceans. Rather, they are more closely related to arachnids (e.g., spiders, mites). Their blood is blue and it clots when bacteria are present, so humans use it to test for dangerous bacteria in injectable drugs, intravenous liquids, and implantable medical devices. "Blue blood" in Spanish indicates nobility.

Atlantic Horseshoe Crabs (*Limulus polyphemus*) play a key role in the ecology of Atlantic Coast estuaries like Delaware Bay. At that 2000 km² Ramsar Wetland of International Importance, they arrive from the ocean depths to spawn in intertidal waters in May and early June. The female digs a nest 10–15 cm deep in soft substrate and lays up to 90,000 eggs over the spawning season. Only 10 of these eggs, on average, will avoid predation and survive to adulthood in 9–12 years. The male that is attached to the female, and often other "satellite" males, fertilize the eggs. Tens of thousands of shorebirds arrive at Delaware Bay at this time. Red Knots, which have declined greatly, gorge on these eggs to put on enough fat to fly directly to their next stop in the Arctic. Remaining fat is used for survival there until the snow melts under unpredictable weather conditions. The knots must also switch their bodies physiologically from migration mode to breeding condition, which also takes considerable energy. When live horseshoe crabs are returned to Delaware Bay after being bled for medicine, many die while others are disoriented and may fail to breed. This human practice, along with harvesting for bait in commercial fishing, threatens their population and that of the Red Knots.

Dennis Paulson

Red Knots and other sandpipers of the genus *Calidris* may be more susceptible than other shorebird species to extinction from changing conditions caused by humans—these species lost much of their all-important genetic variability in so-called "population bottlenecks" during glaciation periods when their population numbers dropped drastically (Baker et al. 2013).

Adapted from NatureServe, HBW, and others.

NatureServe
Conservation Status

Global: ...... Apparently Secure

Canada: .... Vulnerable

US: .... Apparently Secure

Mexico: .... Not Applicable

www.natureserve.org

Surfbird

Scientific: *Calidris virgata*
Français: Bécasseau du ressac
México: Playero Brincaolas
Order: Charadriiformes
Family: Scolopacidae

The Surfbird's former genus name *Aphriza* means "froth" in Greek, referring to its coastal nonbreeding habitat within only several meters of sea foam. Its winter distribution extends from southern Alaska to southern Chile and is one of the longest (17,500 km) and narrowest of any North American breeding bird. If an oil spill were to occur in the South Pacific Ocean, it could easily be carried by the South Pacific Current to much of the west coast of South America. Similarly, an oil spill in the North Pacific Ocean could be carried by the North Pacific Current along much of the west coast of North America. In either case the oil could potentially cover half of the species' global wintering range and thus have a disastrous effect on the global population. This species often associates with the Black Turnstone outside of the breeding season, apparently taking advantage of the latter's alertness and noisy disposition to avoid predators (Paulson 1993; Senner and Mccaffery 1997).

The likelihood of an oil spill in the Surfbird's breeding range is low because the birds are spread out over alpine tundra. They frequent the rocky ridges of mountains that are interspersed with lichen, moss, avens, and heather. When the birds return to their breeding grounds after spring migration, these elevated, darkly colored areas are free of snow because of exposure to intense sunlight and high winds, whereas the lower elevations usually still have snow. Much of this alpine breeding habitat is protected on public lands such as Denali National Park in Alaska and similar areas in the Yukon. Even unprotected areas are likely not threatened because this high-latitude habitat is of little use for resource extraction. This species is consistently recorded in the thousands of individuals at Prince William Sound/Kachemak Bay, Alaska, their last spring migration stopover region.

Larry Master

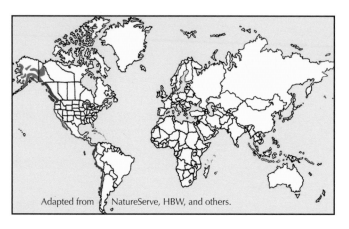

Adapted from NatureServe, HBW, and others.

NatureServe
Conservation Status

Global: .... Secure

Canada: .... Vulnerable

US: .... Secure

Mexico: .... Not Applicable

www.natureserve.org

Stilt Sandpiper

Scientific: *Calidris himantopus*
Français: Bécasseau à échasses
México: Playero Zancón
Order: Charadriiformes
Family: Scolopacidae

The cost of getting high

Carlos Gutierrez

Allen's Rule was proposed by the American zoologist Joel Asaph Allen in 1877. It states that warm-blooded animals from cold climates tend to have shorter limbs than those of similar animals from warm climates. Short limbs in cold climates lose precious heat slowly, whereas long limbs in warm climates dissipate excess heat quickly. Allen's rule often applies within and among similar species.

In a study of 17 closely related shorebird species nesting in the Canadian Arctic, Cartar and Morrison (2005) measured weather harshness during the breeding season over the breeding range of each species and found a significant negative relationship between tarsus length and mean weather harshness. This result suggested that the legs of warm-blooded animals may be shorter in colder environments to take advantage of the ground's wind-dampening boundary layer effect and thus reduce heat loss. This would be one cost of "getting high" in the long-legged Stilt Sandpiper, but there must be other costs that work together to impose upper limits to leg length, e.g., the energy needed to produce long legs, the increased heat loss from long legs, and the increased likelihood of the legs breaking. Presumably the advantage of "getting high" in this species is related to its foraging habits by allowing it to secure its own niche, for example by acquiring food in deeper waters than do other shorebirds.

Experienced Stilt Sandpipers often return to their breeding territory of the previous year. These pairs reduce or skip courtship and territorial displays altogether, perhaps to reduce the risk of predation and/or to breed earlier (Klima and Jehl 2012).

Larry Master

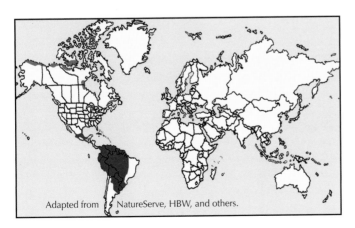

Adapted from NatureServe, HBW, and others.

NatureServe
Conservation Status

Global: .... Secure
Canada: ..... Apparently Secure
US: .... Vulnerable
Mexico: .... Not Applicable

www.natureserve.org

Red-necked Stint

Scientific: *Calidris ruficollis*
Français: Bécasseau à col roux
México: Playero Cuellirojo
Order: Charadriiformes
Family: Scolopacidae

"Peeps" are small (15–19 cm long) sandpipers that are notoriously difficult to distinguish when not in breeding plumage (Cox 2008). The term refers to their vocalizations and small size. In the Old World, peeps are called "stints". Nine peeps/stints exist: the Western, Least, and White-rumped Sandpipers breed only in North America, while the Little, Temminck's, and Long-toed Stints breed only in Eurasia. Most Baird's and Semipalmated Sandpipers breed in North America, while the rest breed in eastern Siberia. Conversely, Red-necked Stints breed almost entirely in northeastern Russia, with only a few breeding in Alaska.

In North America, some birders can identify most shorebird species, except when it comes to the similar little ones; these they are content simply to call "peeps." Others closely examine each peep to distinguish the five common North American species. The elite birder, however, examines each peep in hopes of identifying one of the stints, like the Red-necked Stint. This takes considerable experience and hard work. When I was birding St. Lawrence Island, which belongs to Alaska but is closer to eastern Siberia, someone radioed in a Red-necked Stint to the several independent birding groups present that spring. Twenty of us sprang from our afternoon break in the warm refuge and charged out on 10 noisy all-terrain vehicles. Several dozen birders converged on the find. Most North Americans never get to see this species, much less in its breeding finery. Alaska is excellent for finding Eurasian birds. We were very fortunate to have permission from the local Yupik folks to access the island. Their permission for us to bird the town of Gambell's hillside, which serves as their generations-old cemetery of aboveground coffins exposed to the elements, was particularly gracious.

Captain (Bagsy) Paul Baker

Christian Artuso

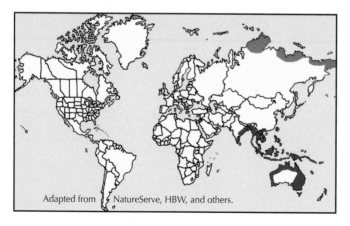

Adapted from NatureServe, HBW, and others.

NatureServe
Conservation Status

Global: .... Secure
Canada: .... Not Applicable
US: .... Vulnerable
Mexico: .... Not Applicable

www.natureserve.org

Sanderling

Scientific: *Calidris alba*
Français: Bécasseau sanderling
México: Playero Blanco
Order: Charadriiformes
Family: Scolopacidae

ARE YOU SEASICK YET?

When foraging on invertebrates on a sandy beach in the nonbreeding season, the Sanderling "haunts the edge of the water, following each retreating wave, and rapidly running back again before the wave's return, threatened every moment to be engulfed in the surf but always just escaping" (Taverner 1919). Its stop-and-start movements are responsible for descriptions of it as a wind-up toy. In spring migration, many Sanderlings stop at Delaware Bay to gorge on Atlantic Horseshoe Crab eggs, though they are not as dependent on them as are Red Knots (see p. 197).

The palest sandpiper in winter plumage, the Sanderling often shows at least some hint of its diagnostic black shoulder in most plumages. In breeding plumage, it is easily mistaken for the smaller Red-necked Stint even though the Sanderling lacks the hind toe of other sandpipers.

Although it has not been studied extensively, the Sanderling's mating system seems to vary among populations and years, ranging from monogamy to serial polyandry. The sexual roles (e.g., which sex incubates) depend on the mating system. In Arctic Canada, for example, one of the parents deserts the "family" at the beginning of incubation. However, if breeding conditions allow the female to lay more clutches, the male incubates the first clutch while the female incubates the second—and if she can lay a third clutch, she is thought to engage a second male. (In the Canadian high Arctic, where snow cover and other environmental conditions vary greatly, laying successive clutches rapidly allows females to raise more young during favorable

David Laliberte

Larry Master

years, though it is presumably more trying for them.) In contrast, both parents of most northeastern Greenland pairs rear the chicks, but only one parent is present at a time (Macwhirter et al. 2002).

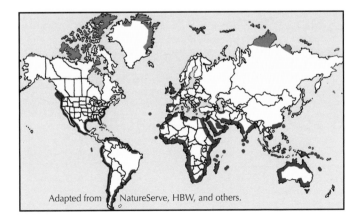

Adapted from NatureServe, HBW, and others.

NatureServe
Conservation Status

Global: .... Secure

Canada: .... Secure

US: .... Apparently Secure

Mexico: .... Not Applicable

www.natureserve.org

Dunlin

Scientific: *Calidris alpina*
Français: Bécasseau variable
México: Playero Dorso Rojo
Order: Charadriiformes
Family: Scolopacidae

A source of concern among researchers working with birds and other animals and plants is the potential impact of their research on the organism being studied. In Dunlins, capture for study reduces the probability of survival over the next year. Within only two hours of capture, Dunlins may lose 14% of their body mass, whereas within four hours they often begin losing muscle mass required for flying. Marking birds for study by applying color dye increases the odds of predation after release unless the birds are held captive until they preen the dye into their feathers, making it less visible. Individuals with installed radio transmitters are more likely to suffer predation within a few days after release (Warnock and Gill 1996). Researchers therefore weigh the proposed benefits of their study against the potential impacts on the birds and determine the optimal conditions under which to collect their data (e.g., extract blood, take measurements, take stomach samples, check nests).

Filming a flock of Dunlins in flight from the ground generally should have little impact on the birds, but it can yield excellent insight into how this species and other flocking birds manage to make abrupt maneuvers "as one bird" (see Semipalmated Sandpiper account, p. 210). Some researchers used to believe that "thought transference" or some kind of electromagnetic communication must be in play. Rather, after Potts (1984) analyzed

Peter Sproule

film of flying Dunlin flocks, he concluded that when one bird in the flock initiates a maneuver, that change in flight pattern spreads through the flock in a "maneuver wave". This wave's spread through the flock starts relatively slowly but soon afterward speeds up to a rate that is three times faster than would be possible if the birds were simply reacting to their immediate neighbors or to a flock leader. Each individual watches birds far down the flock and sets its own time of execution of the maneuver to coincide with the arrival of the wave. Potts called this the "chorus-line hypothesis" because these very fast reaction times seem to be obtained in much the same fashion as they are in a human chorus line.

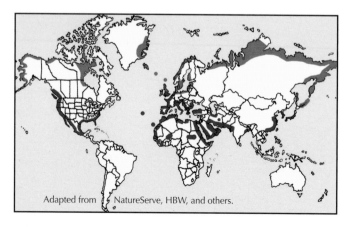

Adapted from NatureServe, HBW, and others.

NatureServe
Conservation Status

Global: .... Secure
Canada: .... Secure
US: .... Secure
Mexico: .... Not Applicable

www.natureserve.org

Rock Sandpiper

Scientific: *Calidris ptilocnemis*
Français: Bécasseau des Aléoutiennes
México: Playero Roquero
Order: Charadriiformes
Family: Scolopacidae

The Rock Sandpiper used to be considered the same species as the Purple Sandpiper. Most taxonomists now consider the two forms as separate species within a superspecies group, meaning that they are particularly close relatives. The former is a western species, whereas the latter is an eastern species, and there is no distributional overlap. However, if they were somehow brought together geographically, it is thought that they would interbreed and produce fertile offspring. This result would prove that the two forms were actually the same species. In fact, given all the human-induced land changes taking place on the Earth, this could happen.

There is often considerable controversy regarding the designation of superspecies, with some taxonomists arguing that the two forms in question are better considered two subspecies within one species. For North American birds, most of the superspecies groups involve two species, but some involve three or more.

When a person moves cautiously toward Rock Sandpipers, the birds run to the top of the boulder they are on and huddle together. Aleutian boys would take advantage of this behavior to procure a meal by hurling a club that would knock the group off the boulder (Bent 1927).

Dennis Paulson

Dennis Paulson

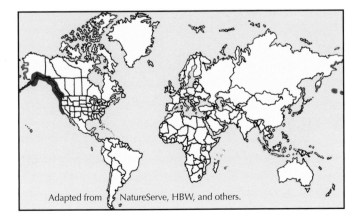

Adapted from NatureServe, HBW, and others.

NatureServe
Conservation Status

Global: .... Secure

Canada: .... Not Applicable

US: .... Secure

Mexico: .... Not Applicable

www.natureserve.org

Purple Sandpiper

Scientific: *Calidris maritima*
Français: Bécasseau violet
México: Playero Oscuro
Order: Charadriiformes
Family: Scolopacidae

I still haven't seen one of these dumpy denizens of the rocky shores, largely because I haven't searched for it in November or December when it is migrating southward long after all the other shorebirds have passed through. In the 1800s, most of the observers (i.e., hunters) thought the Purple Sandpiper was rare because shotguns had been hung up for the season long before these birds arrived. This species' late migration is likely related to the fact that it winters farther north along the Atlantic coast of North America than does any other shorebird, apparently moving only far enough south to avoid particularly harsh conditions. Not surprisingly, it is the most common shorebird in winter on Canada's east coast, Iceland, and Greenland. The reason why this species is able to live in such cold winter conditions as compared with its close relatives is thought to be that mollusks (e.g., sea snails), its primary prey, remain active in winter, unlike invertebrates living in mud or sand.

Whereas some individuals migrate up to 3500 km, others migrate less than 100 km. Some populations breeding in Iceland and southwestern Greenland are thought not to migrate at all because some birds marked during the breeding season have been found close by in winter. In unmarked birds, arrival and departure dates from study sites have also been suggestive, as have measurements of morphological characteristics that aided in separating breeding populations. A complicating factor for this hypothesis, however, is that these two large islands may serve not only as breeding areas for several populations, but also as key migratory staging areas for other populations that breed somewhere else.

David Laliberté

In migration and winter, the Purple Sandpiper frequents rocky coastal shores, but even when it is seen away from the coast, a good place to look for it is on the stones of breakwaters or jetties. It has benefited from the construction of such human structures (Payne and Pierce 2002).

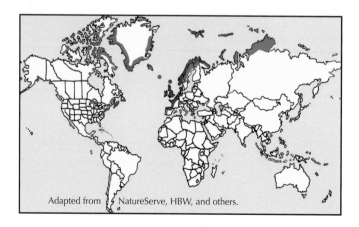

Adapted from NatureServe, HBW, and others.

NatureServe
Conservation Status

Global: .... Secure
Canada:..... Apparently Secure
US: .... Not Applicable
Mexico: .... Not Applicable

www.natureserve.org

Baird's Sandpiper

Scientific: *Calidris bairdii*
Français: Bécasseau de Baird
México: Playero de Baird
Order: Charadriiformes
Family: Scolopacidae

In bird identification, it is often useful to note subtleties about the birds' behavior. For example, Sibley (2001) notes that while they are foraging, shorebirds often segregate by species such that Least Sandpipers are often found close to marsh vegetation, or even within it, whereas Baird's Sandpipers prefer habitats away from water such as the upper parts of beaches. Baird's Sandpipers also prefer inland areas to coastal ones on migration, and their breeding habitat is generally drier than that of closely related shorebirds. They favor well-drained coastal tundra that is exposed to wind, especially sites that become free of snow early in spring.

In one of the costliest egg investments known in female birds, the Baird's Sandpiper lays four eggs that together weigh up to 1.2 times her own weight. Somehow she accomplishes this soon after arriving from migration and despite having little fat left over. Other related unanswered questions are how this species can metabolize energy so quickly during migration, and why it is the only shorebird known to forage regularly in mountains during migration (Moskoff and Montgomerie 2002).

Edgar T. Jones

Larry Master

Adapted from NatureServe, HBW, and others.

NatureServe
Conservation Status

Global: .... Secure
Canada: ..... Apparently Secure
US: .... Apparently Secure
Mexico: .... Not Applicable

www.natureserve.org

Least Sandpiper

Scientific: *Calidris minutilla*
Français: Bécasseau miniscule
México: Playero Diminuto
Order: Charadriiformes
Family: Scolopacidae

The easiest way to confirm a Least Sandpiper identification is to convince yourself that you've seen at *least* one yellow leg of one peep. However, the Least Sandpiper's legs are often muddy and therefore look dark. No problem, just examine some other peeps until you find that yellow leg. The other issue is that two other "peeps", the Long-toed Stint and Temminck's Stint, both Eurasian strays to North America, also have yellow legs. No one will question you if you report a Least Sandpiper based on having seen a yellow leg, but if you post a Long-toed Stint on *e-Bird*, you might create a stir among the birding community. It might therefore be wise to study your bird carefully and build a case for your identification using as many criteria as you can. The professional guide memorizes as many of the criteria as possible for all the species rarely seen in North America (and even potential new species for this region) and examines each bird carefully to pick out that rarity. Your approach thus depends on your priorities.

How closely related are the Least Sandpiper and Long-toed Stint anyway? They are adjacent to each other—according to morphology, plumage, song, and other behavioral characteristics. However, a phylogenetic "supertree" (an amalgamation of existing trees, most of them based on DNA evidence) identified the three most closely related species to the Least Sandpiper as the Semipalmated Sandpiper, Little Stint, and White-rumped Sandpiper (Nebel and Cooper 2008).

Larry Master

The species concept facilitates understanding and communication, but is not a completely accurate portrayal of nature. It's a paradigm, and history notes notable replacements of some by newer ones during "scientific revolutions" (Kuhn 1970)—something to ponder, perhaps, when examining yellow-legged peeps.

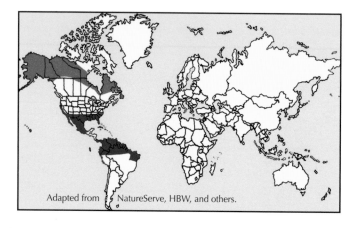

Adapted from NatureServe, HBW, and others.

NatureServe
Conservation Status

Global: .... Secure

Canada: .... Secure

US: .... Secure

Mexico: .... Not Applicable

www.natureserve.org

White-rumped Sandpiper

Scientific: *Calidris fuscicollis*
Français: Bécasseau à croupion blanc
México: Playero Rabadilla Blanca
Order: Charadriiformes
Family: Scolopacidae

The female White-rumped Sandpiper, like the females of numerous other animal species, must deal with a polygynous male. She does most of the work related to raising a particular brood. She selects the nest site in hummocky tundra, constructs the nest, lays the eggs, develops incubation patches, and incubates. She raises only one brood, whereas her mate may sire two or more by mating with other females. The nest cup, which is deep and just wide enough to fit four eggs, is thought to be shaped by the female twirling her body. It is only partially open above, to help conceal the eggs.

How does she occupy herself during the three-week incubation period? Much of her time while awake is spent watching for predators. When one gets too close, she cowers to make herself less visible. She regularly preens her feathers to keep them in good flying condition. She must remember to turn the eggs several times per day to ensure they hatch, and must remember *not* to do so during the last few days before hatching, so the embryos will be properly aligned at hatch time. Passing insects can be good snacks. Sleep time is usually from 9 pm to 5 am, and she gets 20–30% of her day away from the nest in about 25 trips averaging 11 minutes each. The

Larry Master

weather must be monitored so that she flattens herself during rain and faces into strong winds. Hot days make her pant and fluff her feathers to dissipate heat.

Once the eggs hatch she removes the eggshells, but her main jobs are to brood and protect the young. The male is not completely useless, for he displays to her in courtship, then defends the territory until she has laid the eggs. He is then free to court another female, and may even show interest in a female of another closely related species (Parmelee 1992).

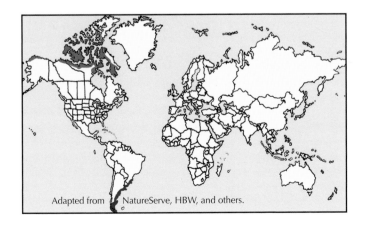

Adapted from NatureServe, HBW, and others.

NatureServe
Conservation Status

Global: .... Secure
Canada: .... Secure
US: .... Vulnerable
Mexico: .... Not Applicable

www.natureserve.org

Buff-breasted Sandpiper

Scientific: *Calidris subruficollis*
Français: Bécasseau roussâtre
México: Playero Ocre
Order: Charadriiformes
Family: Scolopacidae

This species is quiet, the males making courtship *tick* notes that sound like the knocking together of two stones, often while making equally quiet nonvocal buzzing sounds of unknown origin. Both sexes also make other vocal sounds in specific situations.

The Buff-breasted Sandpiper is the only lekking North American shorebird, but it shares this habit with two Eurasian species: the Ruff and the Great Snipe (*Gallinago media*). The Buff-breasted Sandpiper's "exploded" lek mating system refers to the large display territories of individual males (1–4 ha) compared to the tiny 1–2 m² territories of Ruffs. In spring, males disperse over a large area, visiting leks several kilometers apart on successive days, thus visiting numerous leks in one season. They may display at one lek for a day, visit other leks for a few days, then return 1–2 weeks later. The number and identity of males on a given lek change daily. Most leks, which are used year after year, are active for only a few days, the highest number of males seen at a time being 21. Males challenge each other physically and vocally, dancing, doing gymnastics, and making displays, often with open wings. They interfere with each other by sneaking onto a neighbor's territory, acting like a female to avoid detection as a

David Laliberte

male, then copulating with the real female before being detected and chased away by the territorial male. Females may nest several kilometers from the lek where they mated, and receive visits from solitary males, particularly once activity at the leks declines. Some of this solitary activity results in fertilization. Including activity at leks and activity away from leks, about 25% of broods are sired by more than one male. Another phenomenon that occurs before "closing time" is homosexual mounting of resident males by their neighbors, but the adaptiveness of this behavior is thus far unknown (Lanctot and Laredo 1994).

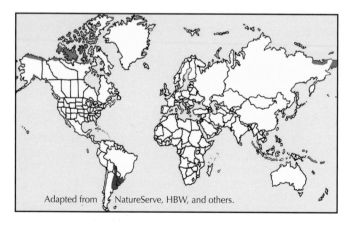

Adapted from NatureServe, HBW, and others.

NatureServe
Conservation Status

Global: ...... Apparently Secure
Canada: .... Vulnerable
US: .... Apparently Secure
Mexico: .... Not Applicable

www.natureserve.org

Pectoral Sandpiper

Scientific: *Calidris melanotos*
Français: Bécasseau à poitrine cendrée
México: Playero Pectoral
Order: Charadriiformes
Family: Scolopacidae

"The potential energy stored up in the small richly colored eggs of this northern sandpiper is almost beyond comprehension....When they are but 30 minutes old, the downy chicks' slight legs carry them over the ground with great rapidity.... In three weeks they are awing and six weeks later they are off on their long journey to the south, crossing mighty mountain ridges and great stretches of land and of sea" (H. W. Brandt, in Bent 1927, 174). "There are some young birds so immature, with threads of yellow down still adhering to the feathers of the head, and altogether weak in appearance, that one can scarcely credit the fact that so soon after being hatched they have actually performed the stupendous journey from the northern extremity of North America to the Buenos-Ayrean pampas" (Hudson 1920, 198).

"Pectoral", meaning breast, refers to the male's fat-laden throat sac that expands and contracts during his courtship display flights—not to the abrupt end of the breast streaks, the key field mark, as I had assumed. When he stands, the skin of his neck hangs like a dewlap. (On his northward spring journey, the fat is deposited between the time he reaches North America and the time he arrives at his breeding area.) Three possible physical mechanisms are postulated for expansion of the throat sac, but bulging of the esophagus won top choice when someone blew

Christian Artuso

air into the esophagus of a dead male, thus causing the same expansion seen during male display, while the air sacs of the neck remained empty of air. The same experiment performed on females yielded no expansion. One of the strangest sounds of the Arctic tundra is the hollow hooting of the displaying live male Pectoral Sandpiper.

The Pectoral Sandpiper is believed to have originated in the New World, later colonizing Siberia, where its breeding range overlaps that of the closely related Sharp-tailed Sandpiper (Farmer et al. 2013).

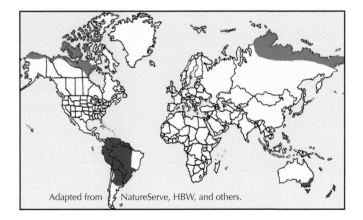

Adapted from NatureServe, HBW, and others.

NatureServe
Conservation Status

Global: .... Secure

Canada: .... Secure

US: .... Secure

Mexico: .... Not Applicable

www.natureserve.org

Semipalmated Sandpiper

Scientific: *Calidris pusilla*
Français: Bécasseau semipalmé
México: Playero Semipalmeado
Order: Charadriiformes
Family: Scolopacidae

"In flight, Semipalmated Sandpipers in flocks, large and small, often move as one bird, twisting and turning with military precision" (Bent 1927, 249). With some luck, you can marvel at flocks of several hundred thousand individuals at key migration staging areas.

Migration researchers are very excited about the useful conservation information they will now be able to obtain readily on this and other declining shorebird species by using the *Motus* Wildlife Tracking System. The main problem with tracking birds has been the weight of the device, including the battery, that must be affixed onto or surgically implanted into the bird. The simplest device is the metal leg band, which is very light, but it yields no information unless retrieved. Other systems work only with larger species or are much more expensive. *Motus* uses small (1.1 cm-long), light (0.3 g), harmless tags attached to the bird's rump that emit a VHF pulse every few seconds and identify the bird and its geographic position at that moment. Thousands of tags can be followed by one researcher on a single radio frequency. An array of towers on the ground automatically picks up the pulses located within a range of 10–75 km of each tower. Some towers store the data for later download, while others transmit the data by internet via wireless dongles. Data are thus obtained on the duration of the birds' stopovers, information that has many uses. For example, changes in site quality can be

Larry Master

detected from changes in stopover times, and the portion of the population stopping at one site that also stops at the next site can be estimated. Another critical piece of the migration puzzle that can now be obtained with better precision is how spring migration behavior affects reproductive success, and how that in turn influences behavior on autumn migration. The various partners of this program plan to expand its network throughout the New World (Paul A. Smith, pers. comm.)

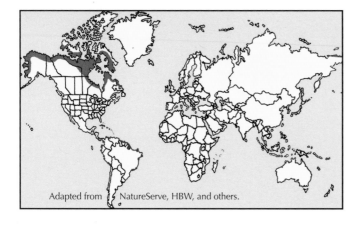

Adapted from NatureServe, HBW, and others.

NatureServe
Conservation Status

Global: .... Secure

Canada: ..... Apparently Secure

US: .... Secure

Mexico: .... Not Applicable

www.natureserve.org

Western Sandpiper

Scientific: *Calidris mauri*
Français: Bécasseau d'Alaska
México: Playero Occidental
Order: Charadriiformes
Family: Scolopacidae

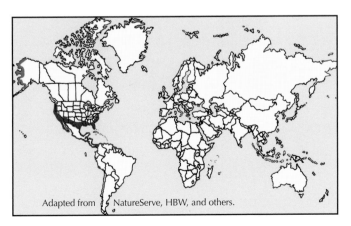

The Western Sandpiper's French name means Alaskan Sandpiper and refers to the fact that most of the global population breeds in Alaska. This species is particularly vulnerable to an environmental disaster because virtually the entire global population of several million birds stops at one site over a few weeks each spring: the Copper River Delta on the central part of Alaska's southern coast. Regardless of the species' abundance, one oil spill could decimate it.

Biofilm refers to any group of microorganisms in which cells adhere to each other on living or non-living surfaces, all held together usually by a matrix produced by the cells. Feeding on biofilm by vertebrates was discovered only recently. Intertidal mudflats and sandflats, where Western Sandpipers love to feed, often have rich biofilm. In fact the many bristles on the sides and tip of this species' tongue seem well adapted for grazing biofilm from these sediment surfaces. At the Fraser Estuary/Boundary Bay system in southwestern British Columbia (a key migration stopover site for Western Sandpipers), samples of their droppings—one of which each bird provided every two minutes of active feeding during migration—showed that biofilm made up 23–53% of their droppings. The importance of biofilm in this species' diet adds a whole new dimension to conservation efforts to manage the few rich stopover sites that it uses during migration (Franks et al. 2014, Jardine et al. 2015), especially when one considers the fact that both biofilm and oil spilled by ships float. Even in waters several kilometers deep, the two meet on the surface.

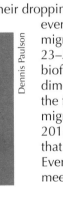
Dennis Paulson

Adapted from NatureServe, HBW, and others.

NatureServe
Conservation Status

Global: .... Secure

Canada: .... Unrankable

US: .... Secure

Mexico: .... Not Applicable

www.natureserve.org

Short-billed Dowitcher

Scientific: *Limnodromus griseus*
Français: Bécassin roux
México: Costurero Pico Corto
Order: Charadriiformes
Family: Scolopacidae

"Oh no, a dowitcher! Hmm, Short-billed (S-b) or Long-billed (L-b)?" It's frustrating when you can record your bird only as "dowitcher sp." (i.e., dowitcher, species unknown). The two species were considered one until 1950 when Frank Pitelka demonstrated that a complicating factor in the *two* species' identification was that the S-b comes in various slightly different looking forms (it is "polytypic").This is besides the fact that the S-b and L-b Dowitchers look very similar. The subtle differences used in distinguishing our two species are treated in painful detail elsewhere, but if you hear the L-b's high *keek* call or the S-b's staccato *tu* in the appropriate contexts (Kaufman 1990), you can skip said detail. The much larger third species, the rare Asian Dowitcher (*L. semipalmatus*), which breeds in eastern Asia (Brazil 2009), is not yet known in North America.

Characteristic of the world's three dowitchers is the fast deep-probing feeding behavior and low mobility compared to the two yellowlegs species and other "sandpipers" of the genus *Tringa*, which make sudden turns. But how do our two dowitchers differ ecologically? Their breeding distributions are quite different, the S-b formerly being called the "Eastern Dowitcher". It molts in winter, whereas the L-b molts in late summer, away from the breeding grounds. In migration, the S-b prefers open coastal saltwater habitats such as mud flats, whereas the L-b frequents inland shallow freshwaters. L-bs migrate later after the breeding season than do most other shorebirds, such that dowitchers seen inland after mid-October are probably this species. In spring in most regions, however, L-bs migrate earlier than do S-bs. You can use these clues to try helping you decide whether your "dowitcher sp." is an S-b or L-b.

Larry Master

Locating the S-b's nest in its muskeg habitats is notoriously difficult (Jehl et al. 2001). Finding the nest of any species requires learning its habits, and this is best mastered by putting in your time in the field.

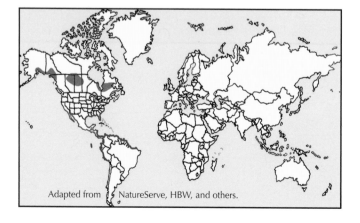

Adapted from NatureServe, HBW, and others.

NatureServe
Conservation Status

Global: .... Secure

Canada: .... Secure

US: .... Secure

Mexico: .... Not Applicable

www.natureserve.org

Long-billed Dowitcher

Scientific: *Limnodromus scolopaceus*
Français: Bécassin à long bec
México: Costurero Pico Largo
Order: Charadriiformes
Family: Scolopacidae

The world's three dowitchers forage using a stitching motion (repeated jabs), a good initial clue in identification, but useless in distinguishing our two species. Dowitchers also probe the sediment, often reaching deeper than the length of the bill while the head is immersed in the shallow water. Herbst corpuscles at the bill tip are loaded with tactile receptors, helping the birds feel for prey (Takekawa and Warnock 2000).

One way to deal with the issue of separating our two dowitchers in the field is to visit Cheyenne Bottoms, Kansas, one of only three remaining large marshes in that state (there were 12 at the time of European colonization) during spring or fall migration. The Long-billed Dowitcher is listed as "abundant" on the site's checklist for both seasons, whereas the Short-billed Dowitcher is "uncommon". (Site checklists are very useful when visiting a birding "hotspot" for the first time because they indicate which species to expect in each season.) When habitat conditions are good, millions of birds (such as an estimated 45% of North America's shorebird population) including many Long-billed Dowitchers, pass through this naturally created land sink in the mixed-grass zone of the Great Plains during spring. Thirty-nine shorebird species occur there. The global Long-billed Dowitcher population is estimated at 500,000, and the highest number seen at the "Bottoms" was an estimated 210,000 in May 1982.

Dennis Paulson

To properly separate our two dowitchers in the field, you might start with Kaufman's (1990) "Advanced Birding" field guide. It devotes eight pages to the topic. The Sibley field guide (2000; 2014) has 18 excellent annotated drawings. Kaufman's second "Advanced Birding" guide (2011) shortened that discussion greatly. Perhaps the topic no longer merited advanced bird identification status because of the publication of other books and articles that introduced the use of features like bill base height, bill curvature, and "loral angle" (see www.surfbirds.com) (Michel Gosselin, pers. comm.).

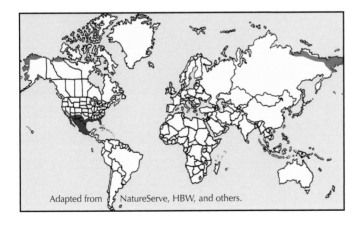

Adapted from NatureServe, HBW, and others.

NatureServe
Conservation Status

Global: .... Secure
Canada: .... Vulnerable
US: .... Secure
Mexico: .... Not Applicable

www.natureserve.org

Wilson's Snipe

Scientific: *Gallinago delicata*
Français: Bécassine de Wilson
México: Agachona Norteamericana
Order: Charadriiformes
Family: Scolopacidae

Birds make different kinds of sounds having various meanings by using a variety of means. In mid- to late-April in Canada's twin-cities National Capital region, when I'm conducting my first of three annual amphibian vocalization surveys half an hour after sunset, I often hear "the weird winnowing sound of the snipe's courtship flight, a tremulous humming sound, loud and penetrating, audible at a long distance" (Bent 1927). This sound is characteristic of very early spring, when the Wood Frogs (*Rana sylvatica*) and Spring Peepers (*Pseudacris crucifer*) are calling, several weeks before the return of most of our passerines in May. It serves as the Wilson Snipe's "song" (territorial advertisement and probably mate attraction), but is made by the rush of air past the spread tail feathers while the rapidly beating wings produce the tremolo effect. Difficult to spot, the "author" of this winnowing is "a mere speck, sweeping across the sky" (Bent 1927).

Non-vocal "songs" are also made by American Woodcocks (during aerial display), Ruffed Grouse (booming), woodpeckers (drumming), herons (bill snapping), Short-eared Owls (wing clapping), and other species. Unlike songs, bird *calls* are vocal sounds made in other contexts,

Netta Smith

such as while feeding, during flight, for aggression, or to express alarm. Other calls are made to keep contact (e.g., between family members) or to beg for food. Birds make other sounds perhaps unwittingly, like the whistling wings of Common Goldeneyes, or the scratching among leaves on the ground by sparrows. They all aid in identification.

"Probably more snipe have been killed by sportsmen than any other game bird.... When the startling cry of the snipe [an altogether different sound from the winnowing described above] arouses the sportsman to instant action he realizes that he is up against a real gamey proposition. He must be a good shot indeed to make a creditable score against such quick erratic flyers" (Bent 1927).

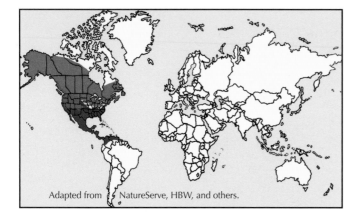

Adapted from NatureServe, HBW, and others.

NatureServe
Conservation Status

Global: .... Secure
Canada: .... Secure
US: .... Secure
Mexico: ... Presumed Extirpated

www.natureserve.org

American Woodcock

Scientific: *Scolopax minor*
Français: Bécasse d'Amérique
México: Chocha del Este
Order: Charadriiformes
Family: Scolopacidae

"The woodcock may be found by those who seek him and know his haunts, but it is only for a short time during the breeding season that he comes out into the open and makes himself conspicuous…. The time to look and listen for it is [in early spring]…. The performance usually begins soon after sunset, as twilight approaches…. The woodcock's nest is usually in some swampy thicket or on the edge of the woods, near an open pasture, field, or clearing; and here in the nearest open space, preferably on some knoll or low hillside within hearing of his sitting mate, the male … entertains her with his thrilling performance. Sometimes, but not always, he struts around on the ground, with tail erect and spread, and with bill pointing downwards and resting on his chest. More often he stands still, or walks about slowly in a normal attitude, producing at intervals of a few seconds two very different notes—a loud, rasping, emphatic *zeeip*—which might be mistaken for the note of the Common Nighthawk, and a soft guttural note, audible at only a short distance, like the croak of a frog or the cluck of a hen. Suddenly he rises, and flies off at a rising angle, circling higher and higher, in increasing spirals, until he looks like a mere speck in the sky, mounting to a height of 60–90 m; during the upward flight he whistles continuously, twittering musical notes…. Then comes his true love song—a loud, musical, three-syllable note given … as he flutters downward, circling, zigzagging, and finally volplaning down to the ground at or near his starting point. He soon begins again on the *zeeip* notes and the whole act is repeated again and again. Sometimes two, or even three, birds may be performing within sight or hearing; occasionally one is seen to drive another away…. This mysterious hermit of the alders… is widely known, but not intimately known" (Bent 1927).

Karen Hanlon

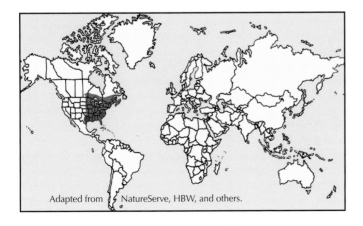

Adapted from NatureServe, HBW, and others.

NatureServe
Conservation Status

Global: .... Secure
Canada: .... Secure
US: .... Secure
Mexico: .... Not Applicable

www.natureserve.org

Wilson's Phalarope

Scientific: *Phalaropus tricolor*
Français: Phalarope de Wilson
México: Falaropo Pico Largo
Order: Charadriiformes
Family: Scolopacidae

"Unlike the other two world-wide phalarope species, the Wilson's Phalarope is a strictly American bird.... It differs from the other two also in being less pelagic and more terrestrial; it is seldom, if ever, seen on the oceans, being a bird of the inland marshes; and it prefers to spend more time walking about on land, or wading in shallow water, than swimming on the water. Hence its bill, neck, and legs are longer, and its feet less lobed. It is a more normal shorebird".... When foraging, "instead of swinging from side to side with a rhythmical motion, as do the Red and Red-necked Phalaropes, this species whirls all the way around. Moreover, it keeps on whirling, and though it pauses for the fraction of a second to inspect its chances, it goes on and on again like an industrious, mad clock....

"Mrs. Wilson wears the breeches and is more inclined to club life than she is to household cares. The case is, however, much more serious than we had at first suspected... for Mrs. Wilson is a bigamist, very usually maintaining two establishments" (Bent 1927, 29–33). Only the male develops brood patches and incubates the eggs (see Northern Jacana account, p. 180) and, as in many other polyandrous animal species, he is smaller and less brightly colored than the female.

After the breeding season, most of the adult Wilson's Phalarope population undertakes a molt migration and spends some time in hypersaline lakes in western North America. The abundant brine shrimp (*Artemia* spp.) and brine flies (*Ephydra* spp.) allow the staging birds to molt quickly

Christian Artuso

and fatten up for migration, while the high water salinity apparently rids the birds of external parasites. Much of the population then flies to saline lakes in the highlands of the central Andes Mountains. For several hours before quitting the staging areas, some adults are too heavy to fly and can be picked up by hand (see Eared Grebe account, p. 87).

A way to increase the flow of fresh water to saline lakes in western North America must be found soon, especially during the current lengthy drought, because they are drying up and becoming too salty (Colwell and Jehl 1994).

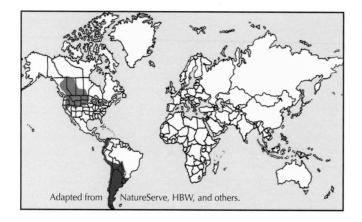

Adapted from NatureServe, HBW, and others.

NatureServe
Conservation Status

Global: .... Secure

Canada: .... Secure

US: .... Secure

Mexico: .... Not Applicable

www.natureserve.org

Red-necked Phalarope

Scientific: *Phalaropus lobatus*
Français: Phalarope à bec étroit
México: Falaropo Cuello Rojo
Order: Charadriiformes
Family: Scolopacidae

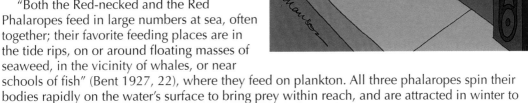

"This is the smallest, the most abundant, and the most widely distributed of the phalaropes; consequently it is the best known" (Bent 1927, 15). "The female … is much more richly colored than the male and possesses all the 'rights' demanded by the most radical reformers" (Nelson 1887, 99). (All three phalarope species are polyandrous.)

"Both the Red-necked and the Red Phalaropes feed in large numbers at sea, often together; their favorite feeding places are in the tide rips, on or around floating masses of seaweed, in the vicinity of whales, or near schools of fish" (Bent 1927, 22), where they feed on plankton. All three phalaropes spin their bodies rapidly on the water's surface to bring prey within reach, and are attracted in winter to parts of the ocean where their prey concentrate near the surface.

Much of the New World population of the Red-necked Phalarope stages on hypersaline lakes in western North America before continuing south to the Humboldt Current off Peru and Ecuador. In eastern North America, millions used to stage in the western Bay of Fundy during autumn, but very few have been seen there in autumn since the 1990s. A similar decline has occurred in the Pacific Ocean off Japan in spring. Have the millions of missing Red-necked

John Hoyt

Phalaropes moved to other unknown staging sites or has a real and massive population reduction occurred? A possible cause of the disappearance of phalaropes from the Deer Island, Maine, area near the mouth of the Bay of Fundy is the greatly reduced abundance of the copepod *Calanus finmarchicus*. The cause of that decline is unknown (Rubega et al. 2000; Brown et al. 2010). Another proposed cause for the Red-necked Phalarope's disappearance from the Bay of Fundy is the 1982/83 El Niño-Southern Oscillation's (ENSO) effect on wintering waters (Nisbet and Veit 2015).

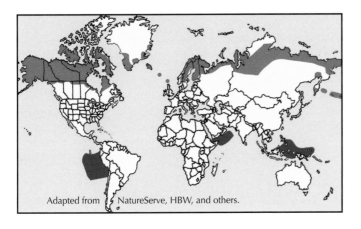

Adapted from NatureServe, HBW, and others.

NatureServe
Conservation Status

Global: ...... Apparently Secure
Canada: ..... Apparently Secure
US: .... Apparently Secure
Mexico: .... Not Applicable

www.natureserve.org

Red Phalarope

Scientific: ***Phalaropus fulicarius***
Français: Phalarope à bec large
México: Falaropo Pico Grueso
Order: Charadriiformes
Family: Scolopacidae

"Phalaropes are active, lively birds in all their movements,…constantly on the move…. They are so much like sandpipers in appearance and in manner of flight that one is always surprised to see them alight on the water…. They float as lightly as corks,… undisturbed by rushing currents or foaming breakers" (Bent 1927, 10).

The Red Phalarope is pelagic in migration and winter, more so than the Red-necked Phalarope. When waters of differing salinity and temperature meet at "oceanic fronts", their unequal densities resist mixing, so the denser water descends, forcing nutrient-rich bottom water to the surface. This creates an oasis together with an associated food chain, from phytoplankton through myriad other organisms up to whales, large fish, and birds. Red Phalaropes thrive there, but also in small ocean oases created by rip tides, and turbulent water crashing over underwater reefs. Stirred-up waters around foraging whales are temporary mobile oases for Red Phalaropes. Shallow areas close to shore provide patches of invertebrate-harboring seaweed, which Red Phalaropes use during storms and during cold periods in spring when their tundra invertebrate prey are inactive.

Most Red Phalaropes breed within 50 km of the ocean on the tundra. During migration near the breeding grounds, Red Phalaropes frequent the edge of the ice pack, leads in sea ice, and

Jenny Erbes

polynyas (permanently ice-free ocean oases). The waters around glaciers and icebergs are also rich in food.

Dramatic population declines like those seen in Red-necked Phalaropes may have occurred without detection in this even more pelagic species, for which monitoring is particularly challenging (Tracy et al. 2002).

"It is a well-known fact that female Red Phalaropes completely ignore their eggs and young ones…. The large, handsome females press their ardent suits against the timid and dull-colored little males" (Bent 1927, 3–4).

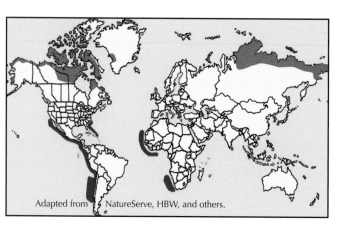

Adapted from NatureServe, HBW, and others.

NatureServe
Conservation Status

Global: .... Secure

Canada: .... Vulnerable

US: .... Secure

Mexico: .... Not Applicable

www.natureserve.org

Literature Cited

Note: A second year shown in parentheses in a reference for a website is the year I accessed the website.

Aebischer, N., and C. Pietri. 1997. "California Quail (*Callipepla californica*)." In *The EBCC Atlas of European Breeding Birds: Their Distribution and Abundance,* E. J. M. Hagemeijer, and M. J. Blair, eds., 216. London: T&AD Poyser.

Ainley, D. G. 1995. "Ashy Storm-Petrel (*Oceanodroma homochroa*)." In *BNA Online,* A. Poole, ed. (2014). Ithaca: Cornell Laboratory of Ornithology. http://bna.birds.cornell.edu/bna/ [Account #185].

Ainley, D. G., D. W. Anderson, and P. R. Kelly. 1981. "Feeding Ecology of Marine Cormorants in Southwestern North America." *Condor* 83:120–131.

Ainley, D. G., and W. T. Everett. 2001. "Black Storm-Petrel (*Oceanodroma melania*)." In *BNA Online,* A. Poole, ed. (2014). Ithaca: Cornell Laboratory of Ornithology. http://bna.birds.cornell.edu/bna/ [Account #577].

Allen, P. E., L. J. Goodrich, and K. L. Bildstein. 1996. "Within- and Among-Year Effects of Cold Fronts on Migrating Raptors at Hawk Mountain, Pennsylvania, 1934–1991." *Auk* 113:329–338.

Alvo, R. 1996. "Common Merganser (*Mergus merganser*)." In *The Breeding Birds of Quebec: Atlas of the Breeding Birds of Southern Quebec,* J. Gauthier, and Y. Aubry, eds., 344–347. Montreal: Association québécoise des groupes d'ornithologues, Province of Quebec Society for the Protection of Birds, and Canadian Wildlife Service, Environment Canada, Quebec Region.

Alvo, R. 2009. "Common Loon, *Gavia immer,* Breeding Success in Relation to Lake pH and Lake Size over 25 Years." *Canadian Field-Naturalist* 123:146–156.

American Ornithologists' Union. 1998. *Check-List of North American Birds,* 7th ed. Washington: American Ornithologists' Union.

American Ornithologists' Union. 2000. "Forty-Second Supplement to the American Ornithologists' Union *Checklist of North American Birds.*" *Auk* 117:847–858.

American Oystercatcher Working Group, E. Nol, and R. C. Humphrey. 2012. "American Oystercatcher (*Haematopus palliatus*)." In *BNA Online,* A. Poole, ed. (2015). Ithaca: Cornell Laboratory of Ornithology. http://bna.birds.cornell.edu/bna/ [Account #82].

Andres, B. A., and G. A. Falxa. 1995. "Black Oystercatcher (*Haematopus bachmani)*." In *BNA Online,* A. Poole, ed. (2015). Ithaca: Cornell Laboratory of Ornithology. http://bna.birds.cornell.edu/bna/ [Account #155].

Anthony, A. W. 1903. "Migration of Richardson's Grouse." *Auk* 20:24–27.

Arterburn, J. W. 2004. "Black-bellied Whistling-Duck (*Dendrocygna autumnalis*)." In *Oklahoma Breeding Bird Atlas,* D. L. Reinking, ed., 60–61. Norman: University of Oklahoma Press.

Austin, J. E., C. M. Custer, and A. D. Afton. 1998. "Lesser Scaup (*Aythya affinis*)." In *BNA Online,* A. Poole, ed. (2014). Ithaca: Cornell Laboratory of Ornithology. http://bna.birds.cornell.edu/bna/ [Account #338].

Austin, J. E., and M. R. Miller. 1995. "Northern Pintail (*Anas acuta*)." In *BNA Online,* A. Poole, ed. (2014). Ithaca: Cornell Laboratory of Ornithology. http://bna.birds.cornell.edu/bna/ [Account #163].

Bailey, E. P. 1993. *Introduction of Foxes to Alaskan Islands: History, Effects on Avifauna, and Eradication.* Resource Publication 193. Washington: United States Department of the Interior, Fish and Wildlife Service.

Bailey, F. M. 1928. *Birds of New Mexico.* United States Bureau of Biological Survey. Santa Fe: New Mexico Department of Game and Fish. Cited in Bent 1932, 75.

Baird, S. F., T. M. Brewer, and R. Ridgway. 1884. *The Water Birds of North America,* Vol. 2. Memoirs of the Museum of Comparative Zoology at Harvard College, Vol. 13. Boston: Little, Brown, and Company.

Baker, A., P. Gonzalez, R.I.G. Morrison, and B. A. Harrington. 2013. "Red Knot (*Calidris canutus*)." In *BNA Online,* A. Poole, ed. (2015). Ithaca: Cornell Laboratory of Ornithology. http://bna.birds.cornell.edu/bna/ [Account #563].

Banks, R. C., R. T. Chesser, C. Cicero, J. L. Dunn, A. W. Kratter, I. J. Lovette, P. C. Rasmussen, J. V. Remsen, Jr., J. D. Rising, D. F. Stotz, and K. Winker. 2008. "Forty-Ninth Supplement to the American Ornithologists' Union *Check-List of North American Birds.*" *Auk* 125:758–768.

Banks, R. C., C. Cicero, J. L. Dunn, A. W. Kratter, P. C. Rasmussen, J. V. Remsen, Jr., J. D. Rising, and D. F. Stotz. 2003. "Forty-Fourth Supplement to the American Ornithologists' Union *Check-List of North American Birds.*" *Auk* 120:923–931.

Banks, R. C., C. Cicero, J. L. Dunn, A. W. Kratter, P. C. Rasmussen, J. V. Remsen, Jr., J. D. Rising, and D. F. Stotz. 2004. "Forty-Fifth Supplement to the American Ornithologists' Union *Check-List of North American Birds.*" *Auk* 121:985–995.

Banks, R. C., C. Cicero, J. L. Dunn, A. W. Kratter, P. C. Rasmussen, J. V. Remsen, Jr., J. D. Rising, and D. F. Stotz. 2006. "Forty-Seventh Supplement to the American Ornithologists' Union *Check-List of North American Birds.*" *Auk* 123:926–936.

Banks, R. C., J. W. Fitzpatrick, T. R. Howell, N. K. Johnson, H. Ouellet, J. V. Remsen, and R. W. Storer. 1985. "Thirty-Fifth Supplement to the American Ornithologists' Union *Check-List of North American Birds.*" *Auk* 102:680–686.

Barbour, D. B., and A. R. DeGrange. 1982. "Communal Roosting in Wintering Hooded Mergansers." *Journal of Field Ornithology* 53:279–280.

Barr, J. F., C. Eberl, and J. W. McIntyre. 2000. "Red-throated Loon (*Gavia stellata*)." In *BNA Online,* A. Poole, ed. (2014). Ithaca: Cornell Laboratory of Ornithology. http://bna.birds.cornell.edu/bna/ [Account #513].

Beauchamp, G. 1992. "Diving Behavior in Surf Scoters and Barrow's Goldeneyes." *Auk* 109:819–827.

Bechard, M. J., C. S. Houston, J. H. Sarasola, and **A. S. England.** 2010. "Swainson's Hawk (*Buteo swainsoni*)." In *BNA Online*, A. Poole, ed. (2014). Ithaca: Cornell Laboratory of Ornithology. http://bna.birds.cornell.edu/bna/ [Account #265].

Bédard, J., A. Nadeau, J-F. Giroux, and **J-P. Savard.** 2008. *Le duvet d'eider: caractéristiques et procédures de récolte.* Ste-Foy: Société Duvetnor Limitée and Canadian Wildlife Service, Environment Canada, Quebec Region, Québec.

Bent, A. C. 1919. *Life Histories of North American Diving Birds: Auk, Grebe, Auklet, Guillemot, Loon, Puffin, Dovekie, Murre, Murrelet.* United States National Museum Bulletin 107. Washington: Smithsonian Institution.

Bent, A. C. 1922. *Life Histories of North American Petrels and Pelicans and their Allies: Gannet, Cormorant, Fulmar, Shearwater, Tropic-bird, Booby, Water-turkey, Pelican, Albatross, Cahow, Petrel, Man-o'-war-bird.* United States National Museum Bulletin 121. Washington: Smithsonian Institution.

Bent, A. C. 1923. *Life Histories of North American Wildfowl,* Part 1: *Duck, Geese, Teals, Mergansers, Eiders, Swans, Scoters, and Others.* United States National Museum Bulletin 126. Washington: Smithsonian Institution.

Bent, A. C. 1925. *Life Histories of North American Wildfowl,* Part 2: *Duck, Geese, Teals, Mergansers, Eiders, Swans, Scoters, and Others.* United States National Museum Bulletin 130. Washington: Smithsonian Institution.

Bent, A. C. 1926. *Life Histories of North American Marsh Birds: Flamingo, Spoonbill, Ibis, Jabiru, Bittern, Heron, Egret, Crane.* United States National Museum Bulletin 135. Washington: Smithsonian Institution.

Bent, A. C. 1927. *Life Histories of North American Shore Birds,* Part 1: *Phalaropes, Snipes, Woodcocks, Sandpipers, Godwits, and others.* United States National Museum Bulletin 142. Washington: Smithsonian Institution.

Bent, A. C. 1929. *Life Histories of North American Shore Birds,* Part 2: *Willets, Oyster Catchers, Tattlers, Plovers, Curlews, and Others.* United States National Museum Bulletin 146. Washington: Smithsonian Institution.

Bent, A. C. 1932. *Life Histories of North American Gallinaceous Birds: Partridge, Bobwhite, Quail, Grouse, Ptarmigan, Prairie Chicken, Pheasant, Turkey, Pigeon, Dove.* United States National Museum Bulletin 162. Washington: Smithsonian Institution.

Bent, A. C. 1937. *Life Histories of North American Birds of Prey,* Part One: *California Condor, Vultures, Kites, Hawks, Eagles, American Osprey.* United States National Museum Bulletin 167. Washington: Smithsonian Institution.

Bergen, B. 1999. "Black-bellied Whistling-Duck (*Dendrocygna autumnalis*)." In *Louisiana Breeding Bird Atlas,* D. A. Wiedenfeld, and M. M. Swan, eds. (2014). Sea Grant College Program. Baton Rouge: Louisiana State University. http://www.manybirds.com/atlas/atlas.htm

Berlanga, H., H. Gómez de Silva, V. M. Vargas-Canales, V. Rodríguez-Contreras, L. A. Sánchez-González, R. Ortega-Álvarez y **R. Calderón-Parra.** (In prep.). Aves de México: Lista actualizada de especies y nombres comunes. CONABIO, México D.F.

Berndt, R. K. 1997. "Gadwall (*Anas strepera*)." In *The EBCC Atlas of European Breeding Birds: Their Distribution and Abundance,* E. J. M. Hagemeijer, and M. J. Blair, eds., 88–89. London: T&AD Poyser.

Bibles, B. D., R. L. Glinski, and **R. R. Johnson.** 2002. "Gray Hawk (*Buteo plagiatus*)." In *BNA Online,* A. Poole, ed. (2014). Ithaca: Cornell Laboratory of Ornithology. http://bna.birds.cornell.edu/bna/ [Account #652].

Bielefeld, R. R., M. G. Brasher, T. E. Moorman, and **P. N. Gray.** 2010. "Mottled Duck (*Anas fulvigula*)." In *BNA Online,* A. Poole, ed. (2014). Ithaca: Cornell Laboratory of Ornithology. http://bna.birds.cornell.edu/bna/ [Account #81].

Bierregaard, R. O., Jr., and **G. M. Kirwan.** 2013. "Hook-billed Kite (*Chondrohierax uncinatus*)." In *del Hoyo et al.* (2015). http://www.hbw.com/node/52955

Bildstein, K. L., and **K. Meyer.** 2000. "Sharp-shinned Hawk (*Accipiter striatus*)." In *BNA Online,* A. Poole, ed. (2014). Ithaca: Cornell Laboratory of Ornithology. http://bna.birds.cornell.edu/bna/ [Account #482].

Blanchan, N. 1899. *Birds that Hunt and are Hunted: Life Histories of One Hundred and Seventy Birds of Prey, Game Birds, and Waterfowls.* New York: Doubleday & McClure Co.

Blohm, R. J. 1981. "Additional Evidence of Egg-Moving Behavior by Female Gadwalls." *Wilson Bulletin* 93:276–277.

Boag, D. A., and **M. A. Schroeder.** 1992. "Spruce Grouse (*Falcipennis canadensis*)." In *BNA Online,* A. Poole, ed. (2014). Ithaca: Cornell Laboratory of Ornithology. http://bna.birds.cornell.edu/bna/ [Account #5].

Bordage, D., and **J-P. Savard.** 2011. "Black Scoter (*Melanitta americana*)." In *BNA Online,* A. Poole, ed. (2014). Ithaca: Cornell Laboratory of Ornithology. http://bna.birds.cornell.edu/bna/ [Account #177].

Boutin, S., C. J. Krebs, A. R. E. Sinclair, and **J. N. M. Smith.** 1986. "Proximate Causes of Losses in a Snowshoe Hare Population." *Canadian Journal of Zoology* 64:606–610.

Braun, C. E., K. Martin, and **L. A. Robb.** 1993. "White-tailed Ptarmigan (*Lagopus leucura*)." In *BNA Online,* A. Poole, ed. (2014). Ithaca: Cornell Laboratory of Ornithology. http://bna.birds.cornell.edu/bna/ [Account #68].

Brazil, M. 2009. *Birds of East Asia.* Princeton: Princeton University Press.

Brennan, L. A. 1999. "Northern Bobwhite (*Colinus virginianis*)." In *BNA Online,* A. Poole, ed. (2014). Ithaca: Cornell Laboratory of Ornithology. http://bna.birds.cornell.edu/bna/ [Account #397].

Brisbin, Jr., I. L., and **T. B. Mowbray.** 2002. "American Coot (*Fulica americana*)." In *BNA Online,* A. Poole, ed. (2014). Ithaca: Cornell Laboratory of Ornithology. http://bna.birds.cornell.edu/bna/ [Account #697a].

Brown, P. W., and **L. H. Fredrickson.** 1997. "White-winged Scoter (*Melanitta fusca*)." In *BNA Online,* A. Poole, ed. (2014). Ithaca: Cornell Laboratory of Ornithology. http://bna.birds.cornell.edu/bna/ [Account #274].

Brown, S., C. Duncan, J. Chardine, and **M. Howe.** 2010. *Red-necked Phalarope Research, Monitoring, and Conservation Plan for the Northeastern U.S. and Maritimes Canada,* Version 1.1. Manomet: Manomet Center for Conservation Sciences.

Brua, R. B. 2002. "Ruddy Duck (*Oxyura jamaicensis*)." In *BNA Online*, A. Poole, ed. (2014). Ithaca: Cornell Laboratory of Ornithology. http://bna.birds.cornell.edu/bna/ [Account #696].

Bryan, D. C. 2002. "Limpkin (*Aramus guarauna*)." In *BNA Online*, A. Poole, ed. (2014). Ithaca: Cornell Laboratory of Ornithology. http://bna.birds.cornell.edu/bna/species/ [Account #627].

Bryant, H. 1861. "On Some Birds Observed in East Florida." *Proceedings of the Boston Society of Natural History* 7:5–21. Cited in Bent 1926, 259.

Buckley, N. J. 1999. "Black Vulture (*Coragyps atratus*)." In *BNA Online*, A. Poole, ed. (2014). Ithaca: Cornell Laboratory of Ornithology. http://bna.birds.cornell.edu/bna/ [Account #411].

Buehler, D. A. 2000. "Bald Eagle (*Haliaeetus leucocephalus*)." In *BNA Online*, A. Poole, ed. (2014). Ithaca: Cornell Laboratory of Ornithology. http://bna.birds.cornell.edu/bna/ [Account #506].

Calkins, J. D., J. C. Hagelin, and D. F. Lott. 1999. "California Quail (*Callipepla californica*)." In *BNA Online*, A. Poole, ed. (2014). Ithaca: Cornell Laboratory of Ornithology. http://bna.birds.cornell.edu/bna/ [Account #473].

Campbell, R. W., N. K. Dawe, I. McTaggart-Cowan, J. M. Cooper, G. W. Kaiser, and M. C. E. McNall. 1990. *The Birds of British Columbia*, Vol. 1, *Nonpasserines: Introduction and Loons Through Waterfowl*. Victoria: Royal British Columbia Museum and Environment Canada, Canadian Wildlife Service.

Carboneras, C., F. Jutglar, A. Bonan, and G. M. Kirwan. 2014. "Northern Fulmar (*Fulmarus glacialis*)." In *del Hoyo et al.* (2014) http://www.hbw.com/node/52514

Cartar, R. V., and R. I. G. Morrison. 2005. "Metabolic Correlates of Leg Length in Breeding Arctic Shorebirds: the Cost of Getting High." *Journal of Biogeography* 32:377–382.

Causey, D. 2002. "Red-faced Cormorant (*Phalacrocorax urile*)." *BNA Online,* A. Poole, ed. (2014). Ithaca: Cornell Laboratory of Ornithology. http://bna.birds.cornell.edu/bna/ [Account #617].

Chapman, F. M. 1891. "On the Birds Observed Near Corpus Christi, Texas, During Parts of March and April, 1891." *Bulletin of the American Museum of Natural History* 3:315–328.

Chapman, F. M. 1908a. *Camps and Cruises of an Ornithologist*. New York: D. Appleton and Company. Cited in Bent 1932, 245.

Chapman, F. M. 1908b. *Camps and Cruises of an Ornithologist*. New York: D. Appleton and Company.

Chesser, R. T., R. C. Banks, C. Cicero, J. L. Dunn, A. W. Kratter, I. J. Lovette, A. G. Navarro-Sigüenza, P. C. Rasmussen, J. V. Remsen, Jr., J. D. Rising, D. F. Stotz, and K. Winker. 2014. "Fifty-fifth Supplement to the American Ornithologists' Union *Check-list of North American Birds*." *Auk: Ornithological Advances* 131:CSi–CSxv.

Chilton, G. 1997. "Labrador Duck (*Camptorhynchus labradorius*)." In *BNA Online*, A. Poole, ed. (2014.) Ithaca: Cornell Laboratory of Ornithology. http://bna.birds.cornell.edu/bna/ [Account #307].

Clay, R. P., A. J. Lesterhuis, and S. Centrón. 2012. "Conservation Plan for the Lesser Yellowlegs (*Tringa flavipes*)", Version 1.0. Manomet: Manomet Center for Conservation Sciences.

Clements, J. F., T. S. Schulenberg, M. J. Iliff, B. L. Sullivan, C. L. Wood, and D. Roberson. 2012. *The eBird/Clements Checklist of Birds of the World*, Version 6.7. http://www.birds.cornell.edu/clementschecklist/download/

Colwell, M. A., and J. R. Jehl, Jr. 1994. Wilson's Phalarope (*Phalaropus tricolor*), In *BNA Online*, A. Poole, ed. (2015). Ithaca: Cornell Laboratory of Ornithology. http://bna.birds.cornell.edu/bna/ [Account #083].

Connelly, J. W., M. W. Gratson, and K. P. Reese. 1998. "Sharp-tailed Grouse (*Tympanuchus phasianellus*)." In *BNA Online*, A. Poole, ed. (2014). Ithaca: Cornell Laboratory of Ornithology. http://bna.birds.cornell.edu/bna/ [Account #354].

Conway, C. J. 1995. "Virginia Rail (*Rallus limicola*)." In *BNA Online*, A. Poole, ed. (2014). Ithaca: Cornell Laboratory of Ornithology. http://bna.birds.cornell.edu/bna/ [Account #173].

Cooper, T. 2007. *King Rail (Rallus elegans) Conservation Plan: A Focal Species Plan Completed by the Midwest Region Division of Migratory Birds*. Version 1.0, US Fish and Wildlife Service. October. (2014). http://www.waterbirdconservation.org/pdfs/King_Rail_Plan_Draft6.pdf

Corbat, C. A., and P. W. Bergstrom. 2000. "Wilson's Plover (*Charadrius wilsonia*)." In *BNA Online*, A. Poole, ed. (2014). Ithaca: Cornell Laboratory of Ornithology. http://bna.birds.cornell.edu/bna/ [Account #516].

Coues, E. 1874. *Birds of the Northwest: a Hand-book of the Ornithology of the Region Drained by the Missouri River and its Tributaries*. United States Geological Survey of the Territories, Miscellaneous Publication 3. Washington: Government Printing Office.

Coulter, M. C., J. A. Rodgers, J. C. Ogden, and F. C. Depkin. 1999. "Wood Stork (*Mycteria americana*)." In *BNA Online*, A. Poole, ed. (2014). Ithaca: Cornell Laboratory of Ornithology. http://bna.birds.cornell.edu/bna/ [Account #409].

Cox, C. 2008. "Identification of North American Peeps: A Different Approach to an Old Problem." *Birding* July/August:32–40.

Cramp, S. 1977. *Handbook of the Birds of Europe, the Middle East, and North Africa: The Birds of the Western Palearctic*, Vol. 1, *Ostrich to Ducks*. Oxford: Oxford University Press.

Cramp, S. 1983. *Handbook of the Birds of Europe, the Middle East, and North Africa: The Birds of the Western Palearctic*, Vol. 3, *Waders to Gulls*. Oxford: Oxford University Press.

Criddle, N. 1917. "The Red-tailed Hawk in Manitoba." *Ottawa Naturalist* 31:74–76.

Cullen, S.A., J. R. Jehl, Jr., and G. L. Nuechterlein. 1999. "Eared Grebe (*Podiceps nigricollis*)." In *BNA Online*, A. Poole, ed. (2014). Ithaca: Cornell Laboratory of Ornithology. http://bna.birds.cornell.edu/bna/ [Account #433].

Curtis, O. E., R. N. Rosenfield, and J. Bielefeldt. 2006. "Cooper's Hawk (*Accipiter cooperii*)." In *BNA Online*, A. Poole, ed. (2014). Ithaca: Cornell Laboratory of Ornithology. http://bna.birds.cornell.edu/bna/ [Account #75].

Dabbert, C. B, G. Pleasant, and S. D. Schemnitz. 2009. "Scaled Quail (*Callipepla squamata*)." In *BNA Online*, A. Poole, ed. (2014). Ithaca: Cornell Laboratory of Ornithology. http://bna.birds.cornell.edu/bna/ [Account #106].

Dale, J. J., and J. E. Thompson. 2001. "Black-bellied Whistling-Duck (*Dendrocygna autumnalis*)." In *BNA Online*, A. Poole, ed. (2014). Ithaca: Cornell Laboratory of Ornithology. http://bna.birds.cornell.edu/bna/ [Account # 578].

Davis, Jr., W. E., and J. Kricher. 2000. "Glossy Ibis (*Plegadis falcinellus*)." In *BNA Online*, A. Poole, ed. (2014). Ithaca: Cornell Laboratory of Ornithology. http://bna.birds.cornell.edu/bna/ [Account # 545].

Davis, Jr., W. E., and J. A. Kushlan. 1994. "Green Heron (*Butorides virescens*)." In *BNA Online*, A. Poole, ed. (2014). Ithaca: Cornell Laboratory of Ornithology. http://bna.birds.cornell.edu/bna/ [Account # 129].

Dawson, W. L. 1909. *The Birds of Washington: a Complete, Scientific, and Popular Account of the 372 Birds Found in the State*, Vol. 2. Seattle: The Occidental Publishing Company.

Dawson, W. L. 1923. *The Birds of California: A Complete, Scientific, and Popular Account of the 580 Species and Subspecies of Birds Found in the State*. San Diego: South Moulton Company. Cited in Bent 1926, 271–272.

Dee Boersma, P., and M. C. Silva. 2001. "Fork-tailed Storm-Petrel (*Oceanodroma furcata*)." In *BNA Online*, A. Poole, ed. (2014). Ithaca: Cornell Laboratory of Ornithology. http://bna.birds.cornell.edu/bna/ [Account # 569].

de la Pena, M. R., and M. Rumboll. 1998. *Birds of Southern South America and Antarctica*. London: HarperCollins Publishers.

del Hoyo, J., Elliott, A., Sargatal, J., Christie, D.A., and de Juana, E. (eds.) 2015. *Handbook of the Birds of the World Alive* (2015). Barcelona: Lynx Edicions. http://www.hbw.com

Delta Waterfowl. 2011. "Al Hochbaum: How He Created the First Duck Research Facility." Press release, Nov. 7, 2011.

Devillers, P., H. Ouellet, É. Benito-Espinal, R. Beudels, R. Cruon, N. David, C. Érard, M. Gosselin, and G. Seutin. 1993. *Noms français des oiseaux du monde: avec les équivalents latins et anglais*. Commission internationale des noms français des oiseaux. Ste-Foy: Éditions MultiMondes inc.

Diamond, A. W., and E. A. Schreiber. 2002. "Magnificent Frigatebird (*Fregata magnificens*)." In *BNA Online*, A. Poole, ed. (2014). Ithaca: Cornell Laboratory of Ornithology. http://bna.birds.cornell.edu/bna/ [Account # 601].

Donovan, L. G., and H. Ouellet. 1993. *Dictionnaire étymologique des noms d'oiseaux du Canada*. Montréal: Guérin.

Drilling, N., R. Titman, and F. Mckinney. 2002. "Mallard (*Anas platyrhynchos*)." In *BNA Online*, A. Poole, ed. (2014). Ithaca: Cornell Laboratory of Ornithology. http://bna.birds.cornell.edu/bna/ [Account # 658].

Dubowy, P. J. 1996. "Northern Shoveler (*Anas clypeata*)." In *BNA Online*, A. Poole, ed. (2014). Ithaca: Cornell Laboratory of Ornithology. http://bna.birds.cornell.edu/bna/ [Account # 217].

Dugger, B. D., and K. M. Dugger. 2002. "Long-billed Curlew (*Numenius americanus*)." In *BNA Online*, A. Poole, ed. (2015). Ithaca: Cornell Laboratory of Ornithology. http://bna.birds.cornell.edu/bna/ [Account # 628].

Dugger, B. D., K. M. Dugger, and L. H. Fredrickson. 2009. "Hooded Merganser (*Lophodytes cucullatus*)." In *BNA Online*, A. Poole, ed. (2014). Ithaca: Cornell Laboratory of Ornithology. http://bna.birds.cornell.edu/bna/ [Account # 98].

Dumas, J. V. 2000. "Roseate Spoonbill (*Platalea ajaja*)." In *BNA Online*, A. Poole, ed. (2014). Ithaca: Cornell Laboratory of Ornithology. http://bna.birds.cornell.edu/bna/ [Account # 490].

Dunk, J. R. 1995. "White-tailed Kite (*Elanus leucurus*)." In *BNA Online*, A. Poole, ed. (2014). Ithaca: Cornell Laboratory of Ornithology. http://bna.birds.cornell.edu/bna/ [Account # 178].

Dunn, E. H., and D. L. Tessaglia. 1994. "Predation on birds at feeders in winter." *Journal of Field Ornithology* 65:8–16.

Dunn, J. L., and J. Alderfer. 2006. *National Geographic Field Guide to the Birds of North America*, 5th ed. Washington: National Geographic Society.

Dunn, J. L., and J. Alderfer. 2011. *National Geographic Field Guide to the Birds of North America*, 6th ed. Washington: National Geographic Society.

Dwyer, J. F., and J. C. Bednarz. 2011. "Harris's Hawk (*Parabuteo unicinctus*)." In *BNA Online*, A. Poole, ed. (2014). Ithaca: Cornell Laboratory of Ornithology. http://bna.birds.cornell.edu/bna/ [Account # 146].

Dykstra, C. R., J. L. Hays, and S. T. Crocoll. 2008. "Red-shouldered Hawk (*Buteo lineatus*)." In *BNA Online*, A. Poole, ed. (2014). Ithaca: Cornell Laboratory of Ornithology. http://bna.birds.cornell.edu/bna/ [Account # 107].

Eadie, J. M., M. L. Mallory, and H. G. Lumsden. 1995. "Common Goldeneye (*Bucephala clangula*)." In *BNA Online*, A. Poole, ed. (2014). Ithaca: Cornell Laboratory of Ornithology. http://bna.birds.cornell.edu/bna/ [Account # 170].

Eadie, J. M., J-P. L. Savard, and M. L. Mallory. 2000. "Barrow's Goldeneye (*Bucephala islandica*)." In *BNA Online*, A. Poole, ed. (2014). Ithaca: Cornell Laboratory of Ornithology. http://bna.birds.cornell.edu/bna/ [Account # 548].

Eaton, S.W. 1992. "Wild Turkey (*Meleagris gallopavo*)." In *BNA Online*, A. Poole, ed. (2013). Ithaca: Cornell Laboratory of Ornithology. http://bna.birds.cornell.edu/bna/ [Account # 22].

eBird. 2012. *eBird: An Online Database of Bird Distribution and Abundance*. Ithaca: eBird. http://www.ebird.org/

Eddleman, W. R., R. E. Flores, and M. Legare. 1994. "Black Rail (*Laterallus jamaicensis*)." In *BNA Online*, A. Poole, ed. (2014). Ithaca: Cornell Laboratory of Ornithology. http://bna.birds.cornell.edu/bna/ [Account #123].

Eggert, A-K. 2014. "Cooperative Breeding in Insects and Vertebrates." In *Oxford Bibliographies in Evolutionary Biology*, J. Losos, ed. (2014). New York: Oxford University Press. www.oxfordbibliographies.com/view/document/obo-9780199941728/obo-9780199941728-0024.xml

Eitniear, J. C. 1999. "Masked Duck (*Nomonyx dominicus*)." In *BNA Online*, A. Poole, ed. (2014). Ithaca: Cornell Laboratory of Ornithology. http://bna.birds.cornell.edu/bna/ [Account # 393].

Elliott, D. G. 1897. *The Gallinaceous Game Birds of North America*. London: Suckling and Company.

Elliott-Smith, E., and S. M. Haig. 2004. "Piping Plover (*Charadrius melodus*)." In *BNA Online*, A. Poole, ed. (2014). Ithaca: Cornell Laboratory of Ornithology. http://bna.birds.cornell.edu/bna/ [Account # 2].

Elphick, C. S., K. C. Parsons, M. Fasola, and L. Mugica. 2010. "Ecology and Conservation of Birds in Rice Fields: a Global Review." *Waterbirds* 33, Special Publication No. 1. Washington: Waterbird Society.

Ely, C. R., and A. X. Dzubin. 1994. "Greater White-fronted Goose (*Anser albifrons*)." In *BNA Online*, A. Poole, ed. (2014). Ithaca: Cornell Laboratory of Ornithology. http://bna.birds.cornell.edu/bna/ [Account #131].

Emlen, S. T., and H. W. Ambrose III. 1970. "Feeding Interactions of Snowy Egrets and Red-breasted Mergansers." *Auk* 87:164–165.

Evers, D. C., J. D. Paruk, J. W. McIntyre, and J. F. Barr. 2010. "Common Loon (*Gavia immer*)." In *BNA Online*, A. Poole, ed. (2014). Ithaca: Cornell Laboratory of Ornithology. http://bna.birds.cornell.edu/bna/ [Account #313].

Farmer, A., R. T. Holmes, and F. A. Pitelka. 2013. "Pectoral Sandpiper (*Calidris melanotos*)." In *BNA Online*, A. Poole, ed. (2015). Ithaca: Cornell Laboratory of Ornithology. http://bna.birds.cornell.edu/bna/ [Account #348].

Farquhar, C. 2009. "White-tailed Hawk (*Buteo albicaudatus*)." In *BNA Online*, A. Poole, ed. (2014). Ithaca: Cornell Laboratory of Ornithology. http://bna.birds.cornell.edu/bna/ [Account #30].

Fisher, B. M. 1975. "American Wigeon Steals Food from Muskrats." *Canadian Field-Naturalist* 89:468.

Fleury, B. E., and T. W. Sherry. 1995. "Long-term Population Trends of Colonial Wading Birds in the Southern United States: the Impact of Crayfish Aquaculture on Louisiana Populations." *Auk* 112:613–632.

Floyd, T. 2008. *Smithsonian Field Guide to the Birds of North America*. New York: HarperCollins Publishers.

Floyd, T., C. S. Elphick, G. Chisholm, K. Mack, R. G. Elston, E. M. Ammon, and J. D. Boone. 2007. *Atlas of the Breeding Birds of Nevada*. Reno and Las Vegas: University of Nevada Press.

Forbush, E. H. 1912. *A History of the Game Birds, Wild-Fowl and Shore Birds of Massachusetts and Adjacent States*. Boston: Massachusetts State Board of Agriculture.

Franks, S., D. B. Lank, and W. H. Wilson. 2014. "Western Sandpiper (*Calidris mauri*)." In *BNA Online*, A. Poole, ed. (2015). Ithaca: Cornell Laboratory of Ornithology. http://bna.birds.cornell.edu/bna/ [Account #90].

Frederick, P. C. 2013. "Tricolored Heron (*Egretta tricolor*)." In *BNA Online*, A. Poole, ed. (2014). Ithaca: Cornell Laboratory of Ornithology. http://bna.birds.cornell.edu/bna/ [Account #306].

Frederick, P. C., and D. Siegel-Causey. 2000. "Anhinga (*Anhinga anhinga*)." In *BNA Online*, A. Poole, ed. (2014). Ithaca: Cornell Laboratory of Ornithology. http://bna.birds.cornell.edu/bna/ [Account #522].

Fredrickson, L. H. 2001. "Steller's Eider (*Polysticta stelleri*)." In *BNA Online*, A. Poole, ed. (2014). Ithaca: Cornell Laboratory of Ornithology. http://bna.birds.cornell.edu/bna/ [Account #571].

Gabrielson, I. N., and F. C. Lincoln. 1959. *The Birds of Alaska*. Harrisburg: The Stackpole Company. Cited in Johnson and Connors 2010, Introduction.

Gale, D., cited in Bendire, C. E. 1892. "Life Histories of North American Birds, with Special Reference to their Breeding Habits and Eggs." *Contributions to Knowledge* 28. Washington: Smithsonian Institution.

Gammonley, J. H. 2012. "Cinnamon Teal (*Anas cyanoptera*)." In *BNA Online*, A. Poole, ed. (2014). Ithaca: Cornell Laboratory of Ornithology. http://bna.birds.cornell.edu/bna/ [Account #209].

Gauthier, G. 1993. "Bufflehead (*Bucephala albeola*)." In *BNA Online*, A. Poole, ed. (2014). Ithaca: Cornell Laboratory of Ornithology. http://bna.birds.cornell.edu/bna/ [Account #67].

Gee, J., D. E. Brown, J. C. Hagelin, M. Taylor, and J. Galloway. 2013. "Gambel's Quail (*Callipepla gambelii*)." In *BNA Online*, A. Poole, ed. (2014). Ithaca: Cornell Laboratory of Ornithology. http://bna.birds.cornell.edu/bna/ [Account #321].

Gerber, B. D., J. F. Dwyer, S. A. Nesbitt, R. C. Drewien, C. D. Littlefield, T. C. Tacha, and P. A. Vohs. 2014. "Sandhill Crane (*Grus canadensis*)." In *BNA Online*, A. Poole, ed. (2014). Ithaca: Cornell Laboratory of Ornithology. http://bna.birds.cornell.edu/bna/ [Account #31].

Gill, R. E., Jr., B. J. Mccaffery, and P. S. Tomkovich. 2002. "Wandering Tattler (*Tringa incana*)." In *BNA Online*, A. Poole, ed. (2015). Ithaca: Cornell Laboratory of Ornithology. http://bna.birds.cornell.edu/bna/ [Account #642].

Gill, R. E., Jr., T. L. Tibbits, D. C. Douglas, C. M. Handel, D. M. Mulcahy, J. C. Gottschalck, N. Warnock, B. J. McCaffery, P. F. Battley, and T. Piersma. 2008. "Extreme Endurance Flights by Landbirds Crossing the Pacific Ocean: Ecological Corridor Rather than Barrier?" (2015.) *Proceedings of the Royal Society* B:1–11. http://alaska.usgs.gov/science/biology/avian_influenza/pdfs/Gill_et_al_2008_Godwit_Migration.pdf

Godfrey, W. E. 1986. *The Birds of Canada*. Ottawa: National Museums of Canada.

Goodrich, L. J., S. T. Crocoll, and S. E. Senner. 2014. "Broad-winged Hawk (*Buteo platypterus*)." In *BNA Online*, A. Poole, ed. (2014). Ithaca: Cornell Laboratory of Ornithology. http://bna.birds.cornell.edu/bna/ [Account #218].

Gordon, S. P. 1915. *Hill Birds of Scotland*. London: Edward Arnold and Company. Cited in Bent 1937, 306–307.

Goss, N. S. 1888. "Feeding Habits of *Pelecanus erythrorhynchus*." *Auk* 5:25.

Gosse, P. H. 1847. *The Birds of Jamaica*. London: John Van Voorst.

Goudie, R. I., G. J. Robertson, and A. Reed. 2000. "Common Eider (*Somateria mollissima*)." In *BNA Online*, A. Poole, ed. (2014). Ithaca: Cornell Laboratory of Ornithology. http://bna.birds.cornell.edu/bna/ [Account #546].

Grant, M. C. 1992. "The Effects of Re-seeding Heathland on Nesting Whimbrel *Numenius phaeopus* in Shetland. 1. Nest Distributions." *Journal of Applied Ecology* 29:501–508.

Grant, M. C., and R. A. Väisänen. 1997. "Whimbrel (*Numenius phaeopus*)." In *The EBCC Atlas of European Breeding Birds: Their Distribution and Abundance*, E. J. M. Hagemeijer, and M. J. Blair, eds., 299. London: T&AD Poyser.

Griscom, L. 1915. "The Little Black Rail on Long Island, N.Y." *Auk* 32:227–228.

Gurnie, J. H. 1913. *The Gannet: a Bird with a History*. London: Witherby and Company.

Gutiérrez, R. J., and D. J. Delehanty. 1999. "Mountain Quail (*Oreortyx pictus*)." In *BNA Online*, A. Poole, ed. (2014). Ithaca: Cornell Laboratory of Ornithology. http://bna.birds.cornell.edu/bna/ [Account #457].

Hagen, C. A., and K. M. Giesen. 2005. "Lesser Prairie-Chicken (*Tympanuchus pallidicinctus*)." In *BNA Online*, A. Poole, ed. (2014). Ithaca: Cornell Laboratory of Ornithology. http://bna.birds.cornell.edu/bna/ [Account #364].

Hamilton, D. J., C. D. Ankney, and R. C. Bailey. 1994. "Predation on Zebra Mussels by Diving Ducks: an Exclosure Study." *Ecology* 75:521–531.

Handel, C. M., and R. E. Gill. 2001. "Black Turnstone (*Arenaria melanocephala*)." In *BNA Online*, A. Poole, ed. (2015). Ithaca: Cornell Laboratory of Ornithology. http://bna.birds.cornell.edu/bna/ [Account # 585].

Hannon, S. J., P. K. Eason, and K. Martin. 1998. "Willow Ptarmigan (*Lagopus lagopus*)." In *BNA Online*, A. Poole, ed. (2014). Ithaca: Cornell Laboratory of Ornithology. http://bna.birds.cornell.edu/bna/ [Account # 369].

Hatch, J. J., K. M. Brown, G. G. Hogan, and R. D. Morris. 2000. "Great Cormorant (*Phalacrocorax carbo*)." In *BNA Online*, A. Poole, ed. (2014). Ithaca: Cornell Laboratory of Ornithology. http://bna.birds.cornell.edu/bna/ [Account # 553].

Hatch, J. J., and D. V. Weseloh. 1999. "Double-crested Cormorant (*Phalacrocorax auritus*)." In *BNA Online*, A. Poole, ed. (2014). Ithaca: Cornell Laboratory of Ornithology. http://bna.birds.cornell.edu/bna/ [Account # 441].

Heath, J. A., P. Frederick, J. A. Kushlan, and K. L. Bildstein. 2009. "White Ibis (*Eudocimus albus*)." In *BNA Online*, A. Poole, ed. (2014). Ithaca: Cornell Laboratory of Ornithology. http://bna.birds.cornell.edu/bna/ [Account # 9].

Helleiner, F. 2013. *For the Birds: Recollections and Rambles*. Brighton: Willow Printing and Publishing Company.

Hepp, G. R., and F. C. Bellrose. 1995. "Wood Duck (*Aix sponsa*)." In *BNA Online*, A. Poole, ed. (2013). Ithaca: Cornell Laboratory of Ornithology. http://bna.birds.cornell.edu/bna/ [Account # 169].

Hobson, K. A. 2013. "Pelagic Cormorant (*Phalacrocorax pelagicus*)." In *BNA Online*, A. Poole, ed. (2014). Ithaca: Cornell Laboratory of Ornithology. http://bna.birds.cornell.edu/bna/ [Account # 282].

Hockey, P., and A. Bonan. 2013. "Family Haematopodidae: Oystercatchers." In *del Hoyo et al.* (2015). http://www.hbw.com/family/oystercatchers-haematopodidae

Hofmann, T., J. W. Chardine, and H. Blokpoel. 1997. "First Breeding Record of Red-breasted Merganser, *Mergus serrator*, on Axel Heiberg Island, Northwest Territories." *Canadian Field-Naturalist* 111:308–309.

Hohman, W. L., and S. A. Lee. 2001. "Fulvous Whistling-Duck (*Dendrocygna bicolor*)." In *BNA Online*, A. Poole, ed. (2014). Ithaca: Cornell Laboratory of Ornithology. http://bna.birds.cornell.edu/bna/ [Account # 562].

Hothem, R. L., B. E. Brussee, and W. E. Davis, Jr. 2010. "Black-crowned Night-Heron (*Nycticorax nycticorax*)." In *BNA Online*, A. Poole, ed. (2014). Ithaca: Cornell Laboratory of Ornithology. http://bna.birds.cornell.edu/bna/ [Account # 74].

Houston, C. S. 2013. "Adventures with White-winged Scoters." *Blue Jay* 71:78–79.

Houston, C. S., C. R. Jackson, and D. E. Bowen, Jr. 2011. "Upland Sandpiper (*Bartramia longicauda*)." In *BNA Online*, A. Poole, ed. (2015). Ithaca: Cornell Laboratory of Ornithology. http://bna.birds.cornell.edu/bna/ [Account # 580].

Houston, D., G. M. Kirwan, and P. Boesman. 2013. "American Black Vulture (*Coragyps atratus*)." In *del Hoyo et al.* (2014). http://www.hbw.com/node/52943

Hudson, W. H. 1920. *The Birds of La Plata*, Vol. 2. London: J. M. Dent & Sons Ltd.

Hughes, B. 1997. "Ruddy Duck (*Oxyura jamaicensis*)." In *The EBCC Atlas of European Breeding Birds: Their Distribution and Abundance*, E. J. M. Hagemeijer, and M. J. Blair, eds., 128. London: T&AD Poyser.

Huntington, C. E., R. G. Butler, and R. A. Mauck. 1996. "Leach's Storm-Petrel (*Oceanodroma leucorhoa*)." In *BNA Online*, A. Poole, ed. (2014). Ithaca: Cornell Laboratory of Ornithology. http://bna.birds.cornell.edu/bna/ [Account # 233].

Jackson, B. J., and J. A. Jackson. 2000. "Killdeer (*Charadrius vociferus*)." In *BNA Online*, A. Poole, ed. (2014). Ithaca: Cornell Laboratory of Ornithology. http://bna.birds.cornell.edu/bna/ [Account # 517].

Jardine, C. B, A. L. Bond, P. J. A. Davidson, R. W. Butler, and T. Kuwae. 2015. "Biofilm Consumption and Variable Diet Composition of Western Sandpipers (*Calidris mauri*) during Migratory Stopover." (2015). *PLoS ONE* 10(4): e0124164. doi:10.1371/journal.pone.0124164.

Jehl, Jr., J. R., J. Klima, and R. E. Harris. 2001. "Short-billed Dowitcher (*Limnodromus griseus*)." In *BNA Online*, A. Poole, ed. (2015). Ithaca: Cornell Laboratory of Ornithology. http://bna.birds.cornell.edu/bna/ [Account # 564].

Jenni, D. A., and T. R. Mace. 1999. "Northern Jacana (*Jacana spinosa*)." In *BNA Online*, A. Poole, ed. (2015). Ithaca: Cornell Laboratory of Ornithology. http://bna.birds.cornell.edu/bna/ [Account # 467].

Johnsgard, P. A., and J. Kear. 1968. "A Review of Parental Carrying of Young by Waterfowl." *The Living Bird* Seventh Annual:89–102. Ithaca: Cornell Laboratory of Ornithology.

Johnson, J. A., M. A. Schroeder, and L. A. Robb. 2011. "Greater Prairie-Chicken (*Tympanuchus cupido*)." In *BNA Online*, A. Poole, ed. (2014). Ithaca: Cornell Laboratory of Ornithology. http://bna.birds.cornell.edu/bna/ [Account # 36].

Johnson, K. 1995. "Green-winged Teal (*Anas crecca*)." In *BNA Online*, A. Poole, ed. (2014). Ithaca: Cornell Laboratory of Ornithology. http://bna.birds.cornell.edu/bna/ [Account # 193].

Johnson, O. W., and P. G. Connors. 2010. "American Golden-Plover (*Pluvialis dominica*)." In *BNA Online*, A. Poole, ed. (2014). Ithaca: Cornell Laboratory of Ornithology. http://bna.birds.cornell.edu/bna/ [Account # 201].

Johnson, R. R., R. L. Glinski, and S. W. Matteson. 2000. "Zone-tailed Hawk (*Buteo albonotatus*)." In *BNA Online*, A. Poole, ed. (2014). Ithaca: Cornell Laboratory of Ornithology. http://bna.birds.cornell.edu/bna/ [Account # 529].

Kaufman, K. 1990. *A Field Guide to Advanced Birding: Birding Challenges and How to Approach Them*. The Peterson Field Guide Series. Boston: Houghton Mifflin Company.

Kaufman, K. 2011. *Kaufman Field Guide to Advanced Birding*. Kaufman Field Guides. New York: Houghton Mifflin Harcourt.

Kaufmann, G. W. 1989. "Breeding Ecology of the Sora (*Porzana carolina*) and the Virginia Rail (*Rallus limicola*)." *Canadian Field-Naturalist* 103:270–282.

Kessel, B, D. A. Rocque, and J. S. Barclay. 2002. "Greater Scaup (*Aythya marila*)." In *BNA Online*, A. Poole, ed. (2014). Ithaca: Cornell Laboratory of Ornithology. http://bna.birds.cornell.edu/bna/ [Account #650].

Keyes, T. S. 2010. "Changes in Georgia's Avifauna since European Settlement." In *The Breeding Bird Atlas of Georgia*, T. M. Schneider, G. Beaton, T. S. Keyes, and N. A. Klaus, eds., 33–45. Athens: University of Georgia Press.

Kilner, R. M. 2006. "Function and Evolution of Color in Young Birds." In *Bird Coloration*, Vol. 2: *Function and Evolution*, G. E. Hill, and K. J. McGraw eds. Cambridge: Harvard University Press.

Kirby, J., and K. Sjöberg. 1997. "Canada Goose (*Branta canadensis*)." In *The EBCC Atlas of European Breeding Birds: Their Distribution and Abundance*, E. J. M. Hagemeijer, and M. J. Blair, eds., 75. London: T&AD Poyser.

Kirk, D. A., and M. J. Mossman. 1998. "Turkey Vulture (*Cathartes aura*)." In *BNA Online*, A. Poole, ed. (2014). Ithaca: Cornell Laboratory of Ornithology. http://bna.birds.cornell.edu/bna/ [Account #339].

Klima, J., and J. R. Jehl, Jr. 2012. "Stilt Sandpiper (*Calidris himantopus*)." In *BNA Online*, A. Poole, ed. (2015). Ithaca: Cornell Laboratory of Ornithology. http://bna.birds.cornell.edu/bna/ [Account #341].

Knopf, F. L., and M. B. Wunder. 2006. "Mountain Plover (*Charadrius montanus*)." In *BNA Online*, A. Poole, ed. (2015). Ithaca: Cornell Laboratory of Ornithology. http://bna.birds.cornell.edu/bna/ [Account #211].

Kochert, M. N., K. Steenhof, C. L. Mcintyre, and E. H. Craig. 2002. "Golden Eagle (*Aquila chrysaetos*)." In BNA Online, A. Poole, ed. (2014). Ithaca: Cornell Laboratory of Ornithology. http://bna.birds.cornell.edu/bna/ [Account #684].

Koes, R. F. 2003. "King Eider (*Somateria spectabilis*)." In *The Birds of Manitoba*, P. Taylor, ed., 117–118. Winnipeg: Manitoba Avian Research Committee, Manitoba Naturalists Society.

Kuhn, T. S. 1970. *The Structure of Scientific Revolutions*, 2nd ed., enlarged. Toward an International Encyclopedia of Unified Science, Foundations of the Unity of Science, Vol. 2, No. 2. Chicago: University of Chicago Press.

Laing, H. M. 1915. "Garoo, Chief Scout of the Prairie." *Outing* 699–710. Cited in Bent 1926, 248.

Lanctot, R. B., and C. D. Laredo. 1994. "Buff-breasted Sandpiper (*Calidris subruficollis*)." In *BNA Online*, A. Poole, ed. (2015). Ithaca: Cornell Laboratory of Ornithology. http://bna.birds.cornell.edu/bna/ [Account #91].

Lee, D. S., and J. C. Haney. 1996. "Manx Shearwater (*Puffinus puffinus*)." In *BNA Online*, A. Poole, ed. (2014). Ithaca: Cornell Laboratory of Ornithology. http://bna.birds.cornell.edu/bna/ [Account #257].

Leschack, C. R., S. K. Mcknight, and G. R. Hepp. 1997. "Gadwall (*Anas strepera*)." In *BNA Online*, A. Poole, ed. (2014). Ithaca: Cornell Laboratory of Ornithology. http://bna.birds.cornell.edu/bna/ [Account #283].

Lewin, V. 1963. "Reproduction and Development of Young in a Population of California Quail." *Condor* 65:249–275.

Limpert, R. J., and S. L. Earnst. 1994. "Tundra Swan (*Cygnus columbianus*)." In *BNA Online*, A. Poole, ed. (2014). Ithaca: Cornell Laboratory of Ornithology. http://bna.birds.cornell.edu/bna/ [Account #89].

Lockwood, M. W., and B. Freeman. 2014. *The Texas Ornithological Society (TOS) Handbook of Texas Birds*, 2nd ed. College Station: Texas A&M University Press.

Loomis, L. M. 1896. "California Water Birds." Numbers 2 and 3, *Proceedings of the California Academy of Sciences, Second Series* 6, pp. 1 and 353. Cited in Bent 1922, 159.

Loss, S. R., T. Will, and P. P. Mara. 2013. "The Impact of Free-Ranging Domestic Cats on Wildlife of the United States." *Nature Communications* 4:1396.

Lowther, P. E., H. D. Douglas III, and C. L. Gratto-Trevor. 2001. "Willet (*Tringa semipalmata*)." In *BNA Online*, A. Poole, ed. (2015). Ithaca: Cornell Laboratory of Ornithology. http://bna.birds.cornell.edu/bna/species/ [Account #579].

Lowther, P. E., and R. T. Paul. 2002. "Reddish Egret (*Egretta rufescens*)." In *BNA Online*, A. Poole, ed. (2014). Ithaca: Cornell Laboratory of Ornithology. http://bna.birds.cornell.edu/bna/ [Account #633].

Lowther, P. E., A. F. Poole, J. P. Gibbs, S. Melvin, and F. A. Reid. 2009. "American Bittern (*Botaurus lentiginosus*)." In *BNA Online*, A. Poole, ed. (2014). Ithaca: Cornell Laboratory of Ornithology. http://bna.birds.cornell.edu/bna [Account #18].

Macwhirter, B., P. Austin-Smith, Jr., and D. Kroodsma. 2002. "Sanderling (*Calidris alba*)." In *BNA Online*, A. Poole, ed. (2015). Ithaca: Cornell Laboratory of Ornithology. http://bna.birds.cornell.edu/bna/ [Account #653].

Maehr, D. S., and H. W. Kale, II. 2005. *Florida's Birds: a Field Guide and Reference*, 2nd. ed. Sarasota, Pineapple Press, Inc.

Mahoney, S. A., and J. R. Jehl, Jr. 1985. "Avoidance of Salt-Loading by a Diving Bird at a Hypersaline and Alkaline Lake: Eared Grebe." *Condor* 87:389–397.

Maley, J. M., and R. T. Brumfield. 2013. "Mitochondrial and Next-Generation Sequence Data used to Infer Phylogenetic Relationships and Species Limits in the Clapper/ King Rail Complex." *Condor* 115:316–329.

Mallory, M. L., S. A. Hatch, and D. N. Nettleship. 2012. "Northern Fulmar (*Fulmarus glacialis*)." In *BNA Online*, A. Poole, ed. (2014). Ithaca: Cornell Laboratory of Ornithology. http://bna.birds.cornell.edu/bna/ [Account #361].

Mallory, M. L., and K. Metz. 1999. "Common Merganser (*Mergus merganser*)." In *BNA Online*, A. Poole, ed. (2014). Ithaca: Cornell Laboratory of Ornithology. http://bna.birds.cornell.edu/bna/ [Account #442].

Manniche, A. L. V. 1910. "The Terrestrial Mammals and Birds of Northeast Greenland: Biological Observations." *Meddelelser om Grønland* 45:1–199. Cited in Bent 1932, 214.

Marchetti, K., and T. Price. 1989. "Differences in the Foraging of Juvenile and Adult Birds: the Importance of Developmental Constraints." *Biological Reviews* 64:51–70.

Marks, J. S., T. L. Tibbitts, R. E. Gill, and B. J. Mccaffery. 2002. "Bristle-thighed Curlew (*Numenius tahitiensis*)." In *BNA Online*, A. Poole, ed. (2015). Ithaca: Cornell Laboratory of Ornithology. http://bna.birds.cornell.edu/bna/ [Account #705].

Marshall, D. B., M. G. Hunter, and A. L. Contreras. 2003. *Birds of Oregon: A General Reference*. Corvallis: Oregon State University Press.

Marti, C., and **E. Lammi.** 1997. "Goosander (*Mergus merganser*)." In *The EBCC Atlas of European Breeding Birds: their Distribution and Abundance*, E. J. M. Hagemeijer, and M. J. Blair, eds., 126–127. London: T&AD Poyser.

Matheu, E., J. del Hoyo, and **E. F. J. Garcia.** 2014. "White-faced Ibis (*Plegadis chihi*)." In *del Hoyo et al. (2015).* http://www.hbw.com/node/52776

McAtee, W. L., and **F. E. L. Beal.** 1912. "Some Common Game, Aquatic, and Rapacious Birds in Relation to Man." *United States Department of Agriculture, Farmers' Bulletin* 497. Washinton: Government Printing Office.

McFarlane, R. 1891. "Notes on and List of Birds and Eggs Collected in Arctic America, 1861–1866." *Proceedings of the United States National Museum* 14:413–446.

McGowan, K. J. 2004. "Introduction: The World of Birds." In *Handbook of Bird Biology*, S. Podulka, R. Rohrbaugh, Jr., and R. Bonney, eds., 1.1–1.13. Ithaca: Cornell Laboratory of Ornithology.

McNeil, R., M. T. Diaz, B. Casanova, and **A. Villeneuve.** 1995. "Trematode parasitism as a possible factor in over-summering of Greater Yellowlegs (*Tringa melanoleuca*)." *Ornitologia Neotropical* 6:57–65.

Melcher, C. P., A. Farmer, and **G. Fernández.** 2010. "Conservation Plan for the Marbled Godwit (*Limosa fedoa*)", Version 1.2. Manomet: Manomet Center for Conservation Science. http://www.whsrn.org/sites/default/files/file/Marbled_Godwit_Conservation_Plan_10_02-28_v1.2.pdf

Melvin, S. M., and **J. P. Gibbs.** 2012. "Sora (*Porzana carolina*)." In *BNA Online*, A. Poole, ed. (2014). Ithaca: Cornell Laboratory of Ornithology. http://bna.birds.cornell.edu/bna/species/ [Account #250].

Merriam-Webster. 2015. *Merriam-Webster Unabridged Style Guide.* http://unabridged.merriam-webster.com

Meyer, K. D. 1995. "Swallow-tailed Kite (*Elanoides forficatus*)." In *BNA Online*, A. Poole, ed. (2014). Ithaca: Cornell Laboratory of Ornithology. http://bna.birds.cornell.edu/bna/ [Account #138].

Miller, J. B. 1996. "Red-breasted Mergansers in an Urban Winter Habitat." *Journal of Field Ornithology* 67:477–483.

Miller, K. E., and **K. D. Meyer.** 2002. "Short-tailed Hawk (*Buteo brachyurus*)." In *BNA Online*, A. Poole, ed. (2014). Ithaca: Cornell Laboratory of Ornithology. http://bna.birds.cornell.edu/bna/ [Account #674].

Mitchell, C. D., and **M. W. Eichholz.** 2010. "Trumpeter Swan (*Cynus buccinator*)." In *BNA Online*, A. Poole, ed. (2014). Ithaca: Cornell Laboratory of Ornithology. http://bna.birds.cornell.edu/bna/ [Account #105].

Mlodinow, S. G. 2005. "Emperor Goose (*Chen canagica*)." In *Birds of Washington: Status and Distribution*, T. R. Wahl, B. Tweit, and S. G. Mlodinow, eds., p. 30. Corvallis: Oregon State University Press.

Montgomerie, R., and **K. Holder.** 2008. "Rock Ptarmigan (*Lagopus muta*)." In *BNA Online*, A. Poole, ed. (2014). Ithaca: Cornell Laboratory of Ornithology. http://bna.birds.cornell.edu/bna/ [Account #51].

Moser, T. J. 2006. *The 2005 North American Trumpeter Swan Survey.* Denver: Division of Migratory Bird Management, US Fish and Wildlife Service.

Moskoff, W., and **R. Montgomerie.** 2002. "Baird's Sandpiper (*Calidris bairdii*)." In *BNA Online*, A. Poole, ed. (2015). Ithaca: Cornell Laboratory of Ornithology. http://bna.birds.cornell.edu/bna/ [Account #661].

Mowbray, T. B. 1999. "American Wigeon (*Anas americana*)." In *BNA Online*, A. Poole, ed. (2014). Ithaca: Cornell Laboratory of Ornithology. http://bna.birds.cornell.edu/bna/ [Account #401].

Mowbray, T. B. 2002a. "Canvasback (*Aythya valisineria*)." In *BNA Online*, A. Poole, ed. (2014). Ithaca: Cornell Laboratory of Ornithology. http://bna.birds.cornell.edu/bna/ [Account #659].

Mowbray, T. B. 2002b. "Northern Gannet (*Morus bassanus*)." In *BNA Online*, A. Poole, ed. (2014). Ithaca: Cornell Laboratory of Ornithology. http://bna.birds.cornell.edu/bna/ [Account #693].

Mowbray, T. B., F. Cooke, and **B. Ganter.** 2000. "Snow Goose (*Chen caerulescens*)." In *BNA Online*, A. Poole, ed. (2014). Ithaca: Cornell Laboratory of Ornithology. http://bna.birds.cornell.edu/bna/ [Account #514].

Mowbray, T. B., C. R. Ely, J. S. Sedinger, and **R. E. Trost.** 2002. "Canada Goose (*Branta canadensis*)." In *BNA Online*, A. Poole, ed. (2014). Ithaca: Cornell Laboratory of Ornithology. http://bna.birds.cornell.edu/bna/ [Account #682].

Murdoch, J. 1885. *Report of the International Polar Expedition to Point Barrow, Alaska*, Part 4: *Natural History.* Cited in Bent 1932, 71.

NatureServe. 2014a. *NatureServe's Central Databases.* Arlington: NatureServe.

NatureServe. 2014b. *NatureServe Explorer: An Online Encyclopedia of Life*, Version 7.1. Arlington: NatureServe. http://www.natureserve.org/explorer

NatureServe. 2015. *NatureServe Explorer: An Online Encyclopedia of Life*, Version 7.1. Arlington: NatureServe. http://www.natureserve.org/explorer

Nebel, S., and **J. M. Cooper.** 2008. "Least Sandpiper (*Calidris minutilla*)." In *BNA Online*, A. Poole, ed. (2015). Ithaca: Cornell Laboratory of Ornithology. http://bna.birds.cornell.edu/bna/ [Account #115].

Nelson, E. W. 1881. "Habits of the Black Brant in the Vicinity of St. Michaels, Alaska." *Bulletin of the Nuttall Ornithological Club* 6:131–138. Cited in Bent 1925, 254.

Nelson, E. W. 1887. *Report Upon Natural History Collections Made in Alaska Between the Years 1877 and 1891.* Arctic Series of Publications Issued in Connection with the Signal Service, US Army, No. 3. Washington: Government Printing Office.

Nettleship, D. N. 2000. "Ruddy Turnstone (*Arenaria interpres*)." In *BNA Online*, A. Poole, ed. (2014). Ithaca: Cornell Laboratory of Ornithology. http://bna.birds.cornell.edu/bna/ [Account #537].

Nisbet, I. C. T., and **R. R. Veit.** 2015. "An Explanation for the Population Crash of Red-necked Phalaropes *Phalaropus lobatus* Staging in the Bay of Fundy in the 1980s." *Marine Ornithology* 43:119–121.

Nol, E., and **M. S. Blanken.** 2014. "Semipalmated Plover (*Charadrius semipalmatus*)." In *BNA Online*, A. Poole, ed. (2014). Ithaca: Cornell Laboratory of Ornithology. http://bna.birds.cornell.edu/bna/ [Account #444].

North American Bird Conservation Initiative Canada. 2012. The State of Canada's Birds. Ottawa: Environment Canada.

Nuechterlein, G. L., and **D. Buitron.** 2002. "Nocturnal Egg Neglect and Prolonged Incubation in the Red-necked Grebe." *Waterbirds* 25:485–491.

Nuechterlein, G. L., and **R. W. Storer.** 1989. "Reverse Mounting in Grebes." *Condor* 91:341–346.

Page, G. W., L. E. Stenzel, G. W. Page, J. S. Warriner, J. C. Warriner, and **P. W. Paton.** 2009. "Snowy Plover (*Charadrius nivosus*)." In *BNA Online*, A. Poole, ed. (2014). Ithaca: Cornell Laboratory of Ornithology. http://bna.birds.cornell.edu/bna/ [Account #154].

Parker, J. W. 1999. "Mississippi Kite (*Ictinia mississippiensis*)", In *BNA Online*, A. Poole, ed. (2014). Ithaca: Cornell Laboratory of Ornithology. http://bna.birds.cornell.edu/bna/ [Account #402].

Parmelee, D. F. 1992. "White-rumped Sandpiper (*Calidris fuscicollis*)." In *BNA Online*, A. Poole, ed. (2015). Ithaca: Cornell Laboratory of Ornithology. http://bna.birds.cornell.edu/bna/ [Account #29].

Parsons, K. C., and **T. L. Master.** 2000. "Snowy Egret (*Egretta thula*)." In *BNA Online*, A. Poole, ed. (2014). Ithaca: Cornell Laboratory of Ornithology. http://bna.birds.cornell.edu/bna/ [Account #489].

Paulson, D. 1993. *Shorebirds of the Pacific Northwest.* Seattle: University of Washington Press.

Payne, L. X., and **E. P. Pierce.** 2002. "Purple Sandpiper (*Calidris maritima*)." In *BNA Online*, A. Poole, ed. (2015). Ithaca: Cornell Laboratory of Ornithology. http://bna.birds.cornell.edu/bna/ [Account #706].

Petersen, M. R., J. B. Grand, and **C. P. Dau.** 2000. "Spectacled Eider (*Somateria fisheri*)." In *BNA Online*, A. Poole, ed. (2014). Ithaca: Cornell Laboratory of Ornithology. http://bna.birds.cornell.edu/bna/ [Account #547].

Petersen, M. R., W. W. Larned, and **D. C. Douglas.** 1999. "At-Sea Distribution of Spectacled Eiders: a 120-Year-Old Mystery Solved." *Auk* 116:1009–1020.

Peterson, M. J. 2000. "Plain Chachalaca (*Ortalis vetula*)." In *BNA Online*, A. Poole, ed. (2014). Ithaca: Cornell Laboratory of Ornithology. http://bna.birds.cornell.edu/bna/ [Account #550].

Peterson, R. T. 1960. *A Field Guide to the Birds of Texas and Adjacent States.* Boston: Houghton Mifflin Company.

Peterson, R. T. 2008. *Peterson Field Guide to Birds of North America.* Boston and New York: Houghton Mifflin Company.

Petrie, S., and **M. L. Schummer.** 2002. "Waterfowl Responses to Zebra Mussels on the Lower Great Lakes." *Birding* (August):346–351. Port Rowan: Bird Studies Canada.

Pettingill, O. S., Jr. 1985. *Ornithology in Laboratory and Field,* 5th ed. Orlando: Academic Press.

Podulka, S., R. W. Rohrbaugh, Jr., and **R. Bonney,** eds. 2004. *Handbook of Bird Biology,* 2nd ed. Ithaca: Cornell Laboratory of Ornithology.

Poole, A. F., L. R. Bevier, C. A. Marantz, and **B. Meanley.** 2005. "King Rail (*Rallus elegans*)." In *BNA Online,* A. Poole, ed. (2014). Ithaca: Cornell Laboratory of Ornithology. http://bna.birds.cornell.edu/bna/ [Account #3].

Poole, A. F., R. O. Bierregaard, and **M. S. Martell.** 2002. "Osprey (*Pandion haliaetus*)." In *BNA Online*, A. Poole, ed. (2014). Ithaca: Cornell Laboratory of Ornithology. http://bna.birds.cornell.edu/bna/ [Account #683].

Potts, W. K. 1984. "The Chorus-Line Hypothesis of Manoeuvre Coordination in Avian Flocks." *Nature* 309:344–345.

Powell, A. N., and **R. S. Suydam.** 2012. "King Eider (*Somateria spectabilis*)." In *BNA Online*, A. Poole, ed. (2014). Ithaca: Cornell Laboratory of Ornithology. http://bna.birds.cornell.edu/bna/ [Account #491].

Pranty, B., J. L. Dunn, S. C. Heinl, A. W. Kratter, P. E. Lehman, M. W. Lockwood, B. Mactavish, and **K. J. Zimmer.** 2008. *ABA Checklist: Birds of the Continental United States and Canada,* 7th ed. Colorado Springs: American Birding Association.

Preston, C. R., and **R. D. Beane.** 2009. "Red-tailed Hawk (*Buteo jamaicensis*)." In *BNA Online*, A. Poole, ed. (2014). Ithaca: Cornell Laboratory of Ornithology. http://bna.birds.cornell.edu/bna/ [Account #52].

Prill, A. G. 1931. "A Land Migration of Coots." *Wilson Bulletin* 43:148–149.

Reed, A., D. H. Ward, D. V. Derksen, and **J. S. Sedinger.** 1998. "Brant (*Branta bernicla*)." In *BNA Online*, A. Poole, ed. (2014). Ithaca: Cornell Laboratory of Ornithology. http://bna.birds.cornell.edu/bna/ [Account #337].

Reed, J. M., L. W. Oring, and **E. M. Gray.** 2013. "Spotted Sandpiper (*Actitis macularius*)." In *BNA Online*, A. Poole, ed. (2015). Ithaca: Cornell Laboratory of Ornithology. http://bna.birds.cornell.edu/bna/ [Account #289].

Ridgely, R. S., T. F. Allnutt, T. Brooks, D. K. McNicol, D. W. Mehlman, B. E. Young, and **J. R. Zook.** 2005. *Digital Distribution Maps of the Birds of the Western Hemisphere,* Version 2.1. Arlington: NatureServe.

Ridgely, R. S., and **P. J. Greenfield.** 2001. *The Birds of Ecuador, Vol. 2: Field Guide.* Ithaca: Comstock Publishing Associates.

Ridgely, R. S., and **J. A. Gwynne, Jr.** 1989. *A Guide to the Birds of Panama, with Costa Rica, Nicaragua, and Honduras,* 2nd ed. Princeton: Princeton University Press.

Robertson, G. J., and **R. I. Goudie.** 1999. "Harlequin Duck (*Histrionicus histrionicus*)." In *BNA Online*, A. Poole, ed. (2013). Ithaca: Cornell Laboratory of Ornithology. http://bna.birds.cornell.edu/bna/ [Account #466].

Robertson, G. J., and **J-P. L. Savard.** 2002. "Long-tailed Duck (*Clangula hyemalis*)." In *BNA Online*, A. Poole, ed. (2014). Ithaca: Cornell Laboratory of Ornithology. http://bna.birds.cornell.edu/bna/ [Account #651].

Rohwer, F. C., W. P. Johnson, and **E. R. Loos.** 2002. "Blue-winged Teal (*Anas discors*)." In *BNA Online*, A. Poole, ed. (2014). Ithaca: Cornell Laboratory of Ornithology. http://bna.birds.cornell.edu/bna/ [Account #625].

Roul, S. 2010. "*Distribution and Status of the Manx Shearwater* (Puffinus puffinus) *on Islands near the Burin Peninsula, Newfoundland.*" Bachelor's Thesis. St. John's: Memorial University of Newfoundland.

Roy, C. L., C. M. Herwig, W. L. Hohman, and **R. T. Eberhardt.** 2012. "Ring-necked Duck (*Aythya collaris*)." In *BNA Online*, A. Poole, ed. (2014). Ithaca: Cornell Laboratory of Ornithology. http://bna.birds.cornell.edu/bna/ [Account #329].

Rubega, M. A., D. Schamel, and **D. M. Tracy.** 2000. "Red-necked Phalarope (*Phalaropus lobatus*." In *BNA Online*, A. Poole, ed. (2015). Ithaca: Cornell Laboratory of Ornithology. http://bna.birds.cornell.edu/bna/ [Account #538].

Rusch, D. H., S. Destefano, M. C. Reynolds, and **D. Lauten.** 2000. "Ruffed Grouse (*Bonasa umbellus*)." In *BNA Online*, A. Poole, ed. (2014). Ithaca: Cornell Laboratory of Ornithology. http://bna.birds.cornell.edu/bna/ [Account #515].

Russell, R. W. 2002. "Arctic Loon (*Gavia pacifica*)." In *BNA Online*, A. Poole, ed. (2014). Ithaca: Cornell Laboratory of Ornithology. http://bna.birds.cornell.edu/bna/ [Account #657].

Ryder, J. P., and **R. T. Alisauskas.** 1995. "Ross's Goose (*Chen rossii*)." In *BNA Online*, A. Poole, ed. (2013). Ithaca: Cornell Laboratory of Ornithology. http://bna.birds.cornell.edu/bna/ [Account #162].

Ryder, R. A. 1998. "White-faced Ibis (*Plegadis chihi*)." In *Colorado Breeding Bird Atlas,* H. E. Kingery, ed., 64–65. Denver: Colorado Bird Atlas Partnership and Colorado Division of Wildlife.

Ryder, R. A., and **D. E. Manry.** 1994. "White-faced Ibis (*Plegadis chihi*)." In *BNA Online*, A. Poole, ed. (2014). Ithaca: Cornell Laboratory of Ornithology. http://bna.birds.cornell.edu/bna/ [Account #130].

Sandys, E., and **T. S. Van Dyke.** 1904. *Upland Game Birds.* New York: The Macmillan Company.

Savard, J-P. L., D. Bordage, and **A. Reed.** 1998. "Surf Scoter (*Melanitta perspicillata*)." In *BNA Online*, A. Poole, ed. (2014). Ithaca: Cornell Laboratory of Ornithology. http://bna.birds.cornell.edu/bna/ [Account #363].

Schindler, D. W. 1998. "A Dim Future for Boreal Waters and Landscapes." *BioScience* 48:157–164.

Schnell, J. H. 1994. "Common Black-Hawk (*Buteogallus anthracinus*)." In *BNA Online*, A. Poole, ed. (2014). Ithaca: Cornell Laboratory of Ornithology. http://bna.birds.cornell.edu/bna/ [Account #122].

Schreiber, E. A., and **R. L. Norton.** 2002. "Brown Booby (*Sula leucogaster*)." In *BNA Online*, A. Poole, ed. (2014). Ithaca: Cornell Laboratory of Ornithology. http://bna.birds.cornell.edu/bna/ [Account #649].

Schroeder, M. A. 2005. "White-tailed Ptarmigan (*Lagopus leucura*)." In *Birds of Washington: Status and Distribution,* T. R. Wahl, B. Tweit, and S. G. Mlodinow, eds., 68–69. Corvallis: Oregon State University Press.

Schroeder, M. A., J. R. Young, and **C. E. Braun.** 1999. "Greater Sage-Grouse (*Centrocercus urophasianus*)." In *BNA Online*, A. Poole, ed. (2014). Ithaca: Cornell Laboratory of Ornithology. http://bna.birds.cornell.edu/bna/ [Account #425].

Schuehammer, A. M., and **S. L. Norris.** 1995. "A Review of the Environmental Impacts of Lead Shotshell Ammunition and Lead Fishing Weights in Canada." *Canadian Wildlife Service Occasional Paper* 88. Ottawa.

Senner, S. E., and **B. J. Mccaffery.** 1997. "Surfbird (*Calidris virgata*)." In *BNA Online*, A. Poole, ed. (2015). Ithaca: Cornell Laboratory of Ornithology. http://bna.birds.cornell.edu/bna/ [Account #266].

Seton, E. T. 1890. Cited in Bent 1932, 292, from E. E. Thompson 1890. "The Birds of Manitoba.*" Proceedings of the US National Museum* 13:457–643.

Shea, R. 1993. "Trumpeter Swan Management." *Henry's Fork Foundation Newsletter* (Fall):4–6.

Shields, M. 1992. "Brown Pelican (*Pelecanus occidentalis*)." In *BNA Online*, A. Poole, ed. (2014). Ithaca: Cornell Laboratory of Ornithology. http://bna.birds.cornell.edu/bna/ [Account #609].

Sibley, D. A. 2000. *The Sibley Guide to Birds.* New York: National Audubon Society, and Alfred A. Knopf.

Sibley, D. A. 2001. *The Sibley Guide to Bird Life and Behavior.* New York: National Audubon Society, and Alfred A. Knopf.

Sibley, D. A. 2014. *The Sibley Guide to Birds*, 2nd edition. New York: Alfred A. Knopf.

Sinclair, P. H., W. A. Nixon, C. D. Eckert, and **N. L. Hughes.** 2003. *Birds of the Yukon Territory.* Vancouver: UBC Press.

Skeel, M. A., and **E. P. Mallory.** 1996. "Whimbrel (*Numenius phaeopus*)." In *BNA Online*, A. Poole, ed. (2015). Ithaca: Cornell Laboratory of Ornithology. http://bna.birds.cornell.edu/bna/ [Account #219].

Skutch, A. F. 1989. *Birds Asleep.* The Corrie Herring Hooks Series 14. Austin: University of Texas Press.

Smith, A. R. 1996. *Atlas of Saskatchewan Birds.* Manley Callin Series 4, Special Publication No. 22. Regina: Environment Canada, and Nature Saskatchewan.

Smith, K. G., S. R. Wittenberg, R. B. Macwhirter, and **K. L. Bildstein.** 2011. "Northern Harrier (*Circus cyaneus*)." In *BNA Online*, A. Poole, ed. (2014). Ithaca: Cornell Laboratory of Ornithology. http://bna.birds.cornell.edu/bna/ [Account #210].

Smith, T. B., and **S. A. Temple.** 1982. "Feeding Habits and Bill Polymorphism in Hook-billed Kites." *Auk* 99:197–207.

Snyder, N. F., and **N. J. Schmitt.** 2002. "California Condor (*Gymnogyps californianus*)." In *BNA Online*, A. Poole, ed. (2014). Ithaca: Cornell Laboratory of Ornithology. http://bna.birds.cornell.edu/bna/ [Account #610].

Squires, J. R., and **R. T. Reynolds.** 1997. "Northern Goshawk (*Accipiter gentilis*)." In *BNA Online*, A. Poole, ed. (2014). Ithaca: Cornell Laboratory of Ornithology. http://bna.birds.cornell.edu/bna/ [Account #298].

Stedman, S. J. 2000. "Horned Grebe (*Podiceps auritus*)." In *BNA Online*, A. Poole, ed. (2014). Ithaca: Cornell Laboratory of Ornithology. http://bna.birds.cornell.edu/bna/ [Account #505].

Stevenson, J. O., and **L. H. Meitzen.** 1946. "Behavior and Food Habits of Sennett's White-tailed Hawk in Texas." *Wilson Bulletin* 58:198–205.

Stiles, F. G., and **A. F. Skutch.** 1989. *A Guide to the Birds of Costa Rica.* Ithaca: Comstock Publishing Associates, a division of Cornell University Press.

Storer, R. W. 2011. "Least Grebe (*Tachybaptus dominicus*)." In *BNA Online*, A. Poole, ed. (2014). Ithaca: Cornell Laboratory of Ornithology. http://bna.birds.cornell.edu/bna/ [Account #24].

Stromberg, M. R. 2000. "Montezuma Quail (*Cyrtonyx montezumae*)." In *BNA Online*, A. Poole, ed. (2014). Ithaca: Cornell Laboratory of Ornithology. http://bna.birds.cornell.edu/bna/ [Account #524].

Suchy, W. J., and **S. H. Anderson.** 1987. "Habitat Suitability Index Models: Northern Pintail." *US Fish and Wildlife Service Biological Report* 82:10.145.

Sykes, Jr., P. W., J. A. Rodgers, Jr., and **R. E. Bennetts.** 1995. "Snail Kite (*Rostrhamus sociabilis*)." In *BNA Online*, A. Poole, ed. (2014). Ithaca: Cornell Laboratory of Ornithology. http://bna.birds.cornell.edu/bna/ [Account #171].

Takekawa, J. Y., and N. Warnock. 2000. "Long-billed Dowitcher (*Limnodromus scolopaceus*)." In *BNA Online*, A. Poole, ed. (2015). Ithaca: Cornell Laboratory of Ornithology. http://bna.birds.cornell.edu/bna/ [Account #493].

Taverner, P. A. 1919. *Birds of Eastern Canada*. Memoir 104, No. 3, Biological Series. Ottawa: Canada Department of Mines, Geological Survey.

Taverner, P. A. 1974. *Birds of Western Canada*. Toronto: Coles Publishing Company Limited.

Taylor, B. 1996. "Sora (*Porzana carolina*)." In *del Hoyo et al.* (2015). http://www.hbw.com/node/53661

Taylor, P. 2003. "White-winged Scoter (*Melanitta fusca*)." In *The Birds of Manitoba*, P. Taylor, ed., 120–121. Winnipeg: Manitoba Avian Research Committee, and Manitoba Naturalists Society.

Telfair, II, R. C. 2006. "Cattle Egret (*Bubulcus ibis*)." In *BNA Online*, A. Poole, ed. (2014). Ithaca: Cornell Laboratory of Ornithology. http://bna.birds.cornell.edu/bna/ [Account #113].

Telfair, II, R. C., and M. L. Morrison. 2005. "Neotropic Cormorant (*Phalacrocorax brasilianus*)." In *BNA Online*, A. Poole, ed. (2014). Ithaca: Cornell Laboratory of Ornithology. http://bna.birds.cornell.edu/bna/ [Account #137].

Terres, J. K. 1991. *The Audubon Society Encyclopedia of North American Birds*. New York: Wings Books.

Than, K. 2006. "Mystery Solved: Why Gorillas Eat Rotting Wood." *Living Science* July 9. http://www.livescience.com/4120-mystery-solved-gorillas-eat-rotting-wood.html

Thompkins, S. 2011. "Rail Hunting Offers Challenge for any Wingshooter." *Houston Chronicle*, September 28. (2015). http://www.chron.com/sports/article/rail-hunting-a-challenge-for-any-wingshooter-2193977.php

Titman, R. D. 1999. "Red-breasted Merganser (*Mergus serrator*)." In *BNA Online*, A. Poole, ed. (2014). Ithaca: Cornell Laboratory of Ornithology. http://bna.birds.cornell.edu/bna/ [Account #443].

Townsend, C. W. 1914. "A Plea for the Conservation of the Eider." *Auk* 31:14–21.

Townsend, M. B. 1910. Letter to A. C. Bent, June 10. Cited in Bent 1923, 113.

Tracy, D. M., D. Schamel, and J. Dale. 2002. "Red Phalarope (*Phalaropus fulicarius*)." In *BNA Online*, A. Poole, ed. (2015). Ithaca: Cornell Laboratory of Ornithology. http://bna.birds.cornell.edu/bna/ [Account #698].

University of Chicago. 2010. *The Chicago Manual of Style*, 16th ed. Chicago: University of Chicago Press.

Vennesland, R. G., and R. W. Butler. 2011. "Great Blue Heron (*Ardea herodias*)." In *BNA Online*, A. Poole, ed. (2014). Ithaca: Cornell Laboratory of Ornithology. http://bna.birds.cornell.edu/bna/ [Account #25].

Vermillion, W. G., and B. C. Wilson. 2009. *Gulf Coast Joint Venture Conservation Planning for Reddish Egret*. Lafayette: Gulf Coast Joint Venture.

Visher, S. S. 1910. "Notes on the Sandhill Crane." *Wilson Bulletin* 22:115.

Walker, B. M., N. R. Senner, C. S. Elphick, and J. Klima. 2011. "Hudsonian Godwit (*Limosa haemastica*)." In *BNA Online*, A. Poole, ed. (2015). Ithaca: Cornell Laboratory of Ornithology. http://bna.birds.cornell.edu/bna/ [Account #629].

Wallace, E. A., and G. E. Wallace. 1998. "Brandt's Cormorant (*Phalacrocorax penicillatus*)." In *BNA Online*, A. Poole, ed. (2014). Ithaca: Cornell Laboratory of Ornithology. http://bna.birds.cornell.edu/bna/ [Account #362].

Walsh, H. M. 1971. *The Outlaw Gunner*. Cambridge: Tidewater Publishers.

Warnock, N. D., and R. E. Gill. 1996. "Dunlin (*Calidris alpina*)." In *BNA Online*, A. Poole, ed. (2015). Ithaca: Cornell Laboratory of Ornithology. http://bna.birds.cornell.edu/bna/ [Account #203].

Watts, B. D. 2011. "Yellow-crowned Night-Heron (*Nyctanassa violacea*)." In *BNA Online*, A. Poole, ed. (2014). Ithaca: Cornell Laboratory of Ornithology. http://bna.birds.cornell.edu/bna/ [Account #161].

Wells, J. V. 2007. *Birder's Conservation Handbook: 100 North American Birds at Risk*. Princeton: Princeton University Press.

Wells, J. V., and P. D. Vickery. 1990. "Willet Nesting in Sphagnum Bog in Maine." *Journal of Field Ornithology* 61:73–75.

West, R. L., and G. K. Hess. 2002. "Purple Gallinule (*Porphyrio martinicus*)." In *BNA Online*, A. Poole, ed. (2014). Ithaca: Cornell Laboratory of Ornithology. http://bna.birds.cornell.edu/bna/ [Account #626].

Wood, W. 1634. *Wood's New-England's Prospect*. Boston: The Prince Society.

Woolfenden, G. E., and W. B. Robertson, Jr. 2006. "Sources and Post-Settlement Changes." In *The Breeding Birds of Florida*, R. F. Noss, ed., 5–70. Special Publication 7. Gainesville: Florida Ornithological Society.

Wormington, A., and J. H. Leach. 1992. "Concentrations of Migrant Diving Ducks at Point Pelee National Park, Ontario, in Response to Invasion of Zebra Mussels, *Dreissena polymorpha*." *Canadian Field-Naturalist* 106:376–380.

Wormworth, J., and K. Mallon. 2006. *Bird Species and Climate Change*. Sydney: World Wildlife Fund for Nature Australia.

Zwickel, F. C., and J. F. Bendell. 2005. "Blue Grouse (*Dendragapus obscurus*)." In *BNA Online*, A. Poole, ed. (2014). Ithaca: Cornell Laboratory of Ornithology. http://bna.birds.cornell.edu/bna/ [Account #15].

Appendix

Species that cannot be, or should not be, protected in North America: regularly occurring nonbreeders, birds of accidental or casual occurrence (reported less frequently than annually), and introduced species (see Introduction pp. 7–8).

ORDER	FAMILY	SCIENTIFIC NAME	ENGLISH NAME
Anseriformes	Anatidae	Anser fabalis	Taiga Bean-Goose
Anseriformes	Anatidae	Anser serrirostris	Tundra Bean-Goose
Anseriformes	Anatidae	Anser brachyrhynchus	Pink-footed Goose
Anseriformes	Anatidae	Anser erythropus	Lesser White-fronted Goose
Anseriformes	Anatidae	Anser anser	Graylag Goose
Anseriformes	Anatidae	Branta leucopsis	Barnacle Goose
Anseriformes	Anatidae	Cygnus olor	Mute Swan
Anseriformes	Anatidae	Cygnus cygnus	Whooper Swan
Anseriformes	Anatidae	Cairina moschata	Muscovy Duck
Anseriformes	Anatidae	Anas falcata	Falcated Duck
Anseriformes	Anatidae	Anas penelope	Eurasian Wigeon
Anseriformes	Anatidae	Anas zonorhyncha	Eastern Spot-billed Duck
Anseriformes	Anatidae	Anas bahamensis	White-cheeked Pintail
Anseriformes	Anatidae	Anas querquedula	Garganey
Anseriformes	Anatidae	Anas formosa	Baikal Teal
Anseriformes	Anatidae	Aythya ferina	Common Pochard
Anseriformes	Anatidae	Aythya fuligula	Tufted Duck
Anseriformes	Anatidae	Mergellus albellus	Smew
Galliformes	Phasianidae	Alectoris chukar	Chukar
Galliformes	Phasianidae	Tetraogallus himalayensis	Himalayan Snowcock
Galliformes	Phasianidae	Perdix perdix	Gray Partridge
Galliformes	Phasianidae	Phasianus colchicus	Ring-necked Pheasant
Galliformes	Phasianidae	Pavo cristatus	Indian Peafowl
Procellariiformes	Diomedeidae	Thalassarche chlororhynchos	Yellow-nosed Albatross
Procellariiformes	Diomedeidae	Thalassarche cauta	White-capped Albatross
Procellariiformes	Diomedeidae	Thalassarche salvini	Salvin's Albatross
Procellariiformes	Diomedeidae	Thalassarche melanophris	Black-browed Albatross
Procellariiformes	Diomedeidae	Phoebetria palpebrata	Light-mantled Albatross
Procellariiformes	Diomedeidae	Diomedea exulans	Wandering Albatross
Procellariiformes	Diomedeidae	Phoebastria immutabilis	Laysan Albatross
Procellariiformes	Diomedeidae	Phoebastria nigripes	Black-footed Albatross
Procellariiformes	Diomedeidae	Phoebastria albatrus	Short-tailed Albatross
Procellariiformes	Procellariidae	Pterodroma macroptera	Great-winged Petrel
Procellariiformes	Procellariidae	Pterodroma solandri	Providence Petrel
Procellariiformes	Procellariidae	Pterodroma arminjoniana	Herald Petrel
Procellariiformes	Procellariidae	Pterodroma ultima	Murphy's Petrel
Procellariiformes	Procellariidae	Pterodroma inexpectata	Mottled Petrel
Procellariiformes	Procellariidae	Pterodroma cahow	Bermuda Petrel
Procellariiformes	Procellariidae	Pterodroma hasitata	Black-capped Petrel
Procellariiformes	Procellariidae	Pterodroma sandwichensis	Hawaiian Petrel
Procellariiformes	Procellariidae	Pterodroma feae	Fea's Petrel
Procellariiformes	Procellariidae	Pterodroma cookii	Cook's Petrel
Procellariiformes	Procellariidae	Pterodroma longirostris	Stejneger's Petrel

FRENCH NAME	MEXICAN NAME	NORTH AMERICA STATUS	G-RANK
Oie des moissons	Ganso Campestre de la Taiga	Regular nonbreeding	G5
Oie de la toundra	Ganso Campestre de la Tundra	Regular nonbreeding	G5
Oie à bec court	Ganso Pico Corto	Accidental or casual	G4
Oie naine	Ganso Chico	Accidental or casual	G4
Oie cendrée	Ganso Común	Accidental or casual	Not tracked
Bernache nonnette	Ganso Cara Blanca	Accidental or casual	G4
Cygne tuberculé	Cisne Vulgar	Introduced	G5
Cygne chanteur	Cisne Cantor	Regular nonbreeding	G5
Canard musqué	Pato Real	Introduced	G4
Canard à faucilles	Pato de Alfanjes	Accidental or casual	G4
Canard siffleur	Pato Silbón	Regular nonbreeding	G5
Canard de Chine	Pato Pico Pinto Oriental	Accidental or casual	G5
Canard des Bahamas	Pato Gargantilla Blanca	Accidental or casual	G4
Sarcelle d'été	Cerceta Cejas Blancas	Accidental or casual	G5
Sarcelle élégante	Cerceta de Baikal	Accidental or casual	G4
Fuligule milouin	Pato Porrón Europeo	Regular nonbreeding	G5
Fuligule morillon	Pato Moñudo	Regular nonbreeding	G5
Harle piette	Mergo Menor	Regular nonbreeding	G4
Perdrix choukar	Perdiz Chukar	Introduced	G5
Tétraogalle de l'Himalaya	Perdigallo Himalayo	Introduced	G5
Perdrix grise	Perdiz Europea	Introduced	G5
Faisan de Colchide	Faisán de Collar	Introduced	G5
Paon blue	Pavo Real	Introduced	G5
Albatros à nez jaune	Albatros Pico Pinto	Accidental or casual	G3
Albatros à cape blanche	Albatros Cautelosa	Accidental or casual	G4
Albatros de Salvin	Albatros de Salvin	Accidental or casual	Not tracked
Albatros à sourcils noirs	Albatros Cejas Negras	Accidental or casual	G3
Albatros fuligineux	Albatros Tiznado	Accidental or casual	G3
Albatros hurleur	Albatros Viajero	Accidental or casual	GNR
Albatros de Laysan	Albatros de Laysan	Regular nonbreeding	G3
Albatros à pieds noirs	Albatros Patas Negras	Regular nonbreeding	G3
Albatros à queue courte	Albatros Rabón	Regular nonbreeding	G1
Pétrel noir	Petrel Alas Grandes	Accidental or casual	G3
Pétrel de Solander	Petrel de Solander	Accidental or casual	Not tracked
Pétrel de la Trinité du Sud	Petrel de Trinidade	Regular nonbreeding	G4
Pétrel de Murphy	Petrel de Murphy	Regular nonbreeding	G2
Pétrel maculé	Petrel Moteado	Regular nonbreeding	G3
Pétrel des Bermudes	Petrel de las Bermudas	Regular nonbreeding	G1
Pétrel diablotin	Petrel Antillano	Regular nonbreeding	G1
Pétrel des Hawaï	Petrel de Hawaii	Regular nonbreeding	G2
Pétrel gongon	Petrel de Cabo Verde	Regular nonbreeding	G1
Pétrel de Cook	Petrel de Cook	Regular nonbreeding	G2
Pétrel de Stejneger	Petrel de Stejneger	Accidental or casual	Not tracked

ORDER	FAMILY	SCIENTIFIC NAME	ENGLISH NAME
Procellariiformes	Procellariidae	Bulweria bulwerii	Bulwer's Petrel
Procellariiformes	Procellariidae	Procellaria aequinoctialis	White-chinned Petrel
Procellariiformes	Procellariidae	Procellaria parkinsoni	Parkinson's Petrel
Procellariiformes	Procellariidae	Calonectris leucomelas	Streaked Shearwater
Procellariiformes	Procellariidae	Calonectris diomedea	Cory's Shearwater
Procellariiformes	Procellariidae	Calonectris edwardsii	Cape Verde Shearwater
Procellariiformes	Procellariidae	Puffinus creatopus	Pink-footed Shearwater
Procellariiformes	Procellariidae	Puffinus carneipes	Flesh-footed Shearwater
Procellariiformes	Procellariidae	Puffinus gravis	Great Shearwater
Procellariiformes	Procellariidae	Puffinus pacificus	Wedge-tailed Shearwater
Procellariiformes	Procellariidae	Puffinus bulleri	Buller's Shearwater
Procellariiformes	Procellariidae	Puffinus griseus	Sooty Shearwater
Procellariiformes	Procellariidae	Puffinus tenuirostris	Short-tailed Shearwater
Procellariiformes	Procellariidae	Puffinus auricularis	Townsend's Shearwater
Procellariiformes	Procellariidae	Puffinus opisthomelas	Black-vented Shearwater
Procellariiformes	Procellariidae	Puffinus lherminieri	Audubon's Shearwater
Procellariiformes	Procellariidae	Puffinus baroli	Barolo Shearwater
Procellariiformes	Hydrobatidae	Oceanites oceanicus	Wilson's Storm-Petrel
Procellariiformes	Hydrobatidae	Pelagodroma marina	White-faced Storm-Petrel
Procellariiformes	Hydrobatidae	Hydrobates pelagicus	European Storm-Petrel
Procellariiformes	Hydrobatidae	Fregetta tropica	Black-bellied Storm-Petrel
Procellariiformes	Hydrobatidae	Oceanodroma hornbyi	Ringed Storm-Petrel
Procellariiformes	Hydrobatidae	Oceanodroma monorhis	Swinhoe's Storm-Petrel
Procellariiformes	Hydrobatidae	Oceanodroma castro	Band-rumped Storm-Petrel
Procellariiformes	Hydrobatidae	Oceanodroma tethys	Wedge-rumped Storm-Petrel
Procellariiformes	Hydrobatidae	Oceanodroma tristrami	Tristram's Storm-Petrel
Procellariiformes	Hydrobatidae	Oceanodroma microsoma	Least Storm-Petrel
Phaethontiformes	Phaethontidae	Phaethon lepturus	White-tailed Tropicbird
Phaethontiformes	Phaethontidae	Phaethon aethereus	Red-billed Tropicbird
Phaethontiformes	Phaethontidae	Phaethon rubricauda	Red-tailed Tropicbird
Ciconiiformes	Ciconiidae	Jabiru mycteria	Jabiru
Suliformes	Fregatidae	Fregata minor	Great Frigatebird
Suliformes	Fregatidae	Fregata ariel	Lesser Frigatebird
Suliformes	Sulidae	Sula dactylatra	Masked Booby
Suliformes	Sulidae	Sula nebouxii	Blue-footed Booby
Suliformes	Sulidae	Sula sula	Red-footed Booby
Pelecaniformes	Ardeidae	Ixobrychus sinensis	Yellow Bittern
Pelecaniformes	Ardeidae	Tigrisoma mexicanum	Bare-throated Tiger-Heron
Pelecaniformes	Ardeidae	Ardea cinerea	Gray Heron
Pelecaniformes	Ardeidae	Mesophoyx intermedia	Intermediate Egret
Pelecaniformes	Ardeidae	Egretta eulophotes	Chinese Egret
Pelecaniformes	Ardeidae	Egretta garzetta	Little Egret
Pelecaniformes	Ardeidae	Egretta gularis	Western Reef-Heron
Pelecaniformes	Ardeidae	Ardeola bacchus	Chinese Pond-Heron
Pelecaniformes	Threskiornithidae	Eudocimus ruber	Scarlet Ibis
Accipitriformes	Accipitridae	Harpagus bidentatus	Double-toothed Kite
Accipitriformes	Accipitridae	Haliaeetus albicilla	White-tailed Eagle
Accipitriformes	Accipitridae	Haliaeetus pelagicus	Steller's Sea-Eagle
Accipitriformes	Accipitridae	Geranospiza caerulescens	Crane Hawk

FRENCH NAME	MEXICAN NAME	NORTH AMERICA STATUS	G-RANK
Pétrel de Bulwer	Petrel de Bulwer	Accidental or casual	G4
Puffin à menton blanc	Petrel Barba Blanca	Accidental or casual	Not tracked
Puffin de Parkinson	Petrel de Parkinson	Accidental or casual	GNR
Puffin leucomèle	Pardela Canosa	Accidental or casual	G3
Puffin cendré	Pardela de Cory	Regular nonbreeding	G5
Puffin du Cap-Vert	Pardela de Cabo Verde	Accidental or casual	GNR
Puffin à pieds roses	Pardela Patas Rosadas	Regular nonbreeding	G3
Puffin à pieds pâles	Pardela Patas Pálidas	Regular nonbreeding	G4
Puffin majeur	Pardela Mayor	Regular nonbreeding	G5
Puffin fouquet	Pardela Cola Cuña	Accidental or casual	G4
Puffin de Buller	Pardela de Buller	Regular nonbreeding	G3
Puffin fuligineux	Pardela Gris	Regular nonbreeding	G5
Puffin à bec grêle	Pardela cola corta	Regular nonbreeding	G5
Puffin de Townsend	Pardela de Revillagigedo	Accidental or casual	G2
Puffin cul-noir	Pardela Mexicana	Regular nonbreeding	G2
Puffin d'Audubon	Pardela de Audubon	Regular nonbreeding	G4
Puffin de Macaronésie	Pardela de Macaronesia	Accidental or casual	Not tracked
Océanite de Wilson	Paíño de Wilson	Regular nonbreeding	G5
Océanite frégate	Paíño Cara Blanca	Regular nonbreeding	G5
Océanite tempête	Paíño Europeo	Accidental or casual	G3
Océanite à ventre noir	Paíño Vientre Negro	Accidental or casual	GNR
Océanite de Hornby	Paíño de Collar	Accidental or casual	GNR
Océanite de Swinhoe	Paíño de Swinhoe	Accidental or casual	GNR
Océanite de Castro	Paíño de Harcourt	Regular nonbreeding	G3
Océanite téthys	Paíño de Galápagos	Accidental or casual	Not tracked
Océanite de Tristram	Paíño de Tristram	Accidental or casual	G3
Océanite minute	Paíño Mínimo	Regular nonbreeding	G3
Phaéton à bec jaune	Rabijunco Cola Blanca	Regular nonbreeding	G5
Phaéton à bec rouge	Rabijunco Pico Rojo	Regular nonbreeding	G5
Phaéton à brins rouges	Rabijunco Cola Roja	Accidental or casual	G4
Jabiru d'Amérique	Cigüeña Jabirú	Accidental or casual	G4
Frégate du Pacifique	Fragata Pelágica	Accidental or casual	G4
Frégate ariel	Fragata Menor	Accidental or casual	G4
Fou masqué	Bobo Enmascarado	Regular nonbreeding	G5
Fou à pieds bleus	Bobo Patas Azules	Accidental or casual	G4
Fou à pieds rouges	Bobo Patas Rojas	Accidental or casual	G5
Blongios de Chine	Avetoro Chino	Accidental or casual	G4
Onoré du Méxique	Garza Tigre Mexicana	Accidental or casual	GNR
Héron cendré	Garza Real	Accidental or casual	GNR
Héron intermédiare	Garza Intermedia	Accidental or casual	G5
Aigrette de Chine	Garza China	Accidental or casual	G3
Aigrette garzette	Garza del Viejo Mundo	Accidental or casual	G5
Aigrette à gorge blanche	Garza Costera	Accidental or casual	G5
Crabier chinois	Garceta China	Accidental or casual	G4
Ibis rouge	Ibis Rojo	Accidental or casual	G4
Milan bidenté	Gavilán Bidentado	Accidental or casual	Not tracked
Pygargue à queue blanche	Águila Cola Blanca	Accidental or casual	G4
Pygargue empereur	Águila de Steller	Accidental or casual	G2
Buse échasse	Gavilán Zancón	Accidental or casual	G4

ORDER	FAMILY	SCIENTIFIC NAME	ENGLISH NAME
Accipitriformes	Accipitridae	Buteo magnirostris	Roadside Hawk
Gruiformes	Rallidae	Crex crex	Corn Crake
Gruiformes	Rallidae	Neocrex erythrops	Paint-billed Crake
Gruiformes	Rallidae	Pardirallus maculatus	Spotted Rail
Gruiformes	Rallidae	Porphyrio flavirostris	Azure Gallinule
Gruiformes	Rallidae	Gallinula chloropus	Common Moorhen
Gruiformes	Rallidae	Fulica atra	Eurasian Coot
Gruiformes	Heliornithidae	Heliornis fulica	Sungrebe
Gruiformes	Gruidae	Grus grus	Common Crane
Charadriiformes	Burhinidae	Burhinus bistriatus	Double-striped Thick-knee
Charadriiformes	Recurvirostridae	Himantopus himantopus	Black-winged Stilt
Charadriiformes	Haematopodidae	Haematopus ostralegus	Eurasian Oystercatcher
Charadriiformes	Charadriidae	Vanellus vanellus	Northern Lapwing
Charadriiformes	Charadriidae	Pluvialis apricaria	European Golden-Plover
Charadriiformes	Charadriidae	Charadrius mongolus	Lesser Sand-Plover
Charadriiformes	Charadriidae	Charadrius leschenaultii	Greater Sand-Plover
Charadriiformes	Charadriidae	Charadrius collaris	Collared Plover
Charadriiformes	Charadriidae	Charadrius dubius	Little Ringed Plover
Charadriiformes	Scolopacidae	Xenus cinereus	Terek Sandpiper
Charadriiformes	Scolopacidae	Actitis hypoleucos	Common Sandpiper
Charadriiformes	Scolopacidae	Tringa ochropus	Green Sandpiper
Charadriiformes	Scolopacidae	Tringa brevipes	Gray-tailed Tattler
Charadriiformes	Scolopacidae	Tringa erythropus	Spotted Redshank
Charadriiformes	Scolopacidae	Tringa nebularia	Common Greenshank
Charadriiformes	Scolopacidae	Tringa stagnatilis	Marsh Sandpiper
Charadriiformes	Scolopacidae	Tringa glareola	Wood Sandpiper
Charadriiformes	Scolopacidae	Tringa totanus	Common Redshank
Charadriiformes	Scolopacidae	Numenius minutus	Little Curlew
Charadriiformes	Scolopacidae	Numenius madagascariensis	Far Eastern Curlew
Charadriiformes	Scolopacidae	Numenius tenuirostris	Slender-billed Curlew
Charadriiformes	Scolopacidae	Numenius arquata	Eurasian Curlew
Charadriiformes	Scolopacidae	Limosa limosa	Black-tailed Godwit
Charadriiformes	Scolopacidae	Calidris tenuirostris	Great Knot
Charadriiformes	Scolopacidae	Calidris pugnax	Ruff
Charadriiformes	Scolopacidae	Calidris falcinellus	Broad-billed Sandpiper
Charadriiformes	Scolopacidae	Calidris acuminata	Sharp-tailed Sandpiper
Charadriiformes	Scolopacidae	Calidris ferruginea	Curlew Sandpiper
Charadriiformes	Scolopacidae	Calidris temminckii	Temminck's Stint
Charadriiformes	Scolopacidae	Calidris subminuta	Long-toed Stint
Charadriiformes	Scolopacidae	Calidris pygmea	Spoon-billed Sandpiper
Charadriiformes	Scolopacidae	Calidris minuta	Little Stint
Charadriiformes	Scolopacidae	Lymnocryptes minimus	Jack Snipe
Charadriiformes	Scolopacidae	Gallinago gallinago	Common Snipe
Charadriiformes	Scolopacidae	Gallinago stenura	Pin-tailed Snipe
Charadriiformes	Scolopacidae	Gallinago solitaria	Solitary Snipe
Charadriiformes	Scolopacidae	Scolopax rusticola	Eurasian Woodcock
Charadriiformes	Glareolidae	Glareola maldivarum	Oriental Pratincole

FRENCH NAME	MEXICAN NAME	NORTH AMERICA STATUS	G-RANK
Buse à gros bec	Aguililla Caminera	Accidental or casual	G5
Râle des genêts	Polluela Euroasiática	Accidental or casual	G5
Râle à bec peint	Polluela Pico Pinto	Accidental or casual	G4
Râle tacheté	Rascón Pinto	Accidental or casual	G4
Talève favorite	Gallineta Celeste	Accidental or casual	Not tracked
Gallinule poule-d'eau	Gallineta del Viejo Mundo	Accidental or casual	Not tracked
Foulque macroule	Gallareta Euroasiática	Accidental or casual	G5
Grébifoulque d'Amérique	Pájaro Cantil	Accidental or casual	GNR
Grue cendrée	Grulla Euroasiática	Accidental or casual	G4
Oedicnème bistrié	Alcaraván Americano	Accidental or casual	G4
Échasse blanche	Candelero del Viejo Mundo	Accidental or casual	G4
Huîtrier pie	Ostrero Euroasiático	Accidental or casual	G5
Vanneau huppé	Avefría Europea	Accidental or casual	G5
Pluvier doré	Chorlo Dorado Europeo	Accidental or casual	G4
Pluvier de Mongolie	Chorlo Mongol Menor	Regular nonbreeding	G4
Pluvier de Leschenault	Chorlo Mongol Mayor	Accidental or casual	GNR
Pluvier d'Azara	Chorlo de Collar	Accidental or casual	G5
Pluvier petit-gravelot	Chorlo Anillado Menor	Accidental or casual	G5
Chevalier bargette	Playero Picopando	Regular nonbreeding	G5
Chevalier guignette	Playero Euroasiático	Regular nonbreeding	G5
Chevalier cul-blanc	Playero Andarríos Mayor	Accidental or casual	G5
Chevalier de Sibérie	Playero Siberiano	Regular nonbreeding	G4
Chevalier arlequin	Patarroja Negra	Accidental or casual	G5
Chevalier aboyeur	Pataverde Mayor	Regular nonbreeding	G5
Chevalier stagnatile	Patamarilla Euroasiática	Accidental or casual	G5
Chevalier sylvain	Playero Andarríos	Regular nonbreeding	G5
Chevalier gambette	Patarroja Común	Accidental or casual	G5
Courlis nain	Zarapito Chico	Accidental or casual	G3
Courlis de Sibérie	Zarapito Siberiano	Accidental or casual	G4
Courlis à bec grêle	Zarapito Pico Fino	Accidental or casual	G2
Courlis cendré	Zarapito Real	Accidental or casual	G5
Barge à queue noire	Picopando Cola Negra	Regular nonbreeding	G5
Becasseau de l'Anadyr	Playero Grande	Accidental or casual	G4
Combattant varié	Combatiente	Regular nonbreeding	G5
Bécasseau falcinelle	Playerito Agachona	Accidental or casual	G4
Bécasseau à queue pointue	Playero Siberiano	Regular nonbreeding	G5
Bécasseau cocorli	Playero Zarapito	Regular nonbreeding	G5
Bécasseau de Temminck	Playerito de Temminck	Regular nonbreeding	G5
Bécasseau à longs doigts	Playerito Dedos Largos	Regular nonbreeding	G4
Bécasseau spatule	Playerito Pico de Cuchara	Accidental or casual	G2
Bécasseau minute	Playerito Menor	Accidental or casual	G5
Bécassine sourde	Agachona Chica	Accidental or casual	G5
Bécassine des marais	Agachona Común Euroasiática	Regular nonbreeding	G5
Bécassine à queue pointue	Agachona Cola Aguda	Accidental or casual	G4
Bécassine solitaire	Agachona Solitaria	Accidental or casual	GNR
Bécasse des bois	Chocha Euroasiática	Accidental or casual	G5
Glaréole orientale	Glareola de Oriente	Accidental or casual	G3

Contributors

recently migrated to the US east coast where she is discovering new birds, wildlife, and islands to explore and photograph. **Frances Alvo**, a music student at Bishop's University, Quebec, spends her summers working at a nature camp near Algonquin Park, Ontario. She began birding at a young age with her father Jack, and enjoys accompanying him and Uncle Rob on birding trips, be they to Point Pelee or Utah's canyons. **Investment professional **Jack Alvo** was introduced to birding by his brother Rob at Point Pelee in 1983. Jack lives in Toronto with his wife and three children. He loves the outdoors (landscapes, canoeing, and hiking), rarely traveling without his bins. Jack is also very active in folk music communities, where he plays guitar and sings. **Well-traveled **Christian Artuso**'s fascination with birds began as a youngster in Australia, and he has since photographed over 3000 species. His PhD at the University of Manitoba examined how Eastern Screech-Owls are affected by human population density. He coordinates the Manitoba Breeding Bird Atlas for Bird Studies Canada. http://artusobirds.blogspot.ca/ | http://artusophotos.com **Born in Frome, England, **Vaughan Ashby** soon became fascinated by birds. He worked as an assistant bank manager for 24 years. In 1991 he founded www.birdfinders.co.uk, one of the largest European birdwatching companies, offering tours to every continent. He is treasurer of Portland Bird Observatory and edited the Dorset Bird Report. **Neurologist **Juan Bahamon** sees immense suffering at work, so he relishes the balance offered by taking artistic bird-portraits. He loves scouting a bird by foot or boat, learning its behavior, slowly approaching it while monitoring changing light conditions, selecting the best lens, and patiently waiting for flattering poses. Or he crawls in mud, hides in hot blinds, or stands in place for hours. Juan follows the teachings of master bird photographers such as Arthur Morris, Charles Glatzer, E. J. Peiker, and Greg Downing. Juan's work has been exhibited in medical buildings, art galleries, medical magazines, calendars, and greeting cards. He was a moderator for the Avian Forum of Nature Photographers Online Magazine and won the editors' pick in November 2002. www.birdsofcorpuschristi.com **Captain (Bagsy) Paul Baker** lives in England, a full-time nomad of no fixed abode. His global wanderings began at age 15 with the Royal Navy, followed by the Merchant Marine, then the offshore oil industry. Having visited 147 countries, he still backpacks the world. Please see http://bagsy-thecaptainslog.blogspot.com/ and pass it on. **Giff Beaton** is a pilot from Marietta, Georgia. He studies, photographs, leads tours, and gives presentations on birds and insects. Publications include: *Birding Georgia, Birds of Georgia, Birds of Kennesaw Mountain, The Breeding Bird Atlas of Georgia* (2010), the *Annotated Checklist of Georgia Birds*, and *Dragonflies and Damselflies of Georgia and the Southeast*. www.giffbeaton.com **A retired communications engineer and lifetime amateur naturalist, **Gord Belyea** maintains a Tree Swallow nest-box trail just south of Ottawa, Canada. He loves observing and photographing birds. Gord and his partner Ann are avid members of the Ottawa Field-Naturalists' Club and spend winters birding in Texas, New Mexico, and Arizona. **Gavin Bieber** has lived and traveled in numerous countries. After studying biology and environmental studies, he taught bird identification and has been involved in numerous ornithological studies. Gavin now works as a guide for WINGS/Sunbird and serves on the Board of Directors of the Tucson Audubon Society. http://wingsbirds.com/leaders/gavin-bieber/ **Todd Black** co-founded Utah State University's Community Based Conservation Program (http://www.utahcbcp.org), for which he trapped, counted, and photographed sage-grouse for conservation. As Wildlife Manager for Desert Western Ranches, Todd oversees wildlife monitoring and environmental lessee/conservation partnership programs on 400,000 ha of land in the western US and western Canada (http://www.deseretlandandlivestock.com). **Right after physician **Carol Blackard** almost bumped into a breeding-plumaged Black-crowned Night-Heron perched above her exercise path, she bought a field guide. Birds extend her appreciation of the wonder of the human body to that of the natural world. Photographing and conserving birds were natural sequels. www.carolblackardphotography.com **Donald Bleitz** was born in Los Angeles in 1915. He invented a remotely controlled camera shutter that could work at 1/100,000th of a second. He photographed more than 600 species of North American birds and planned to publish a detailed book on the birds of North America using his photos of birds paired with plants from their habitats, but did not finish before he died in 1986. However, he published many articles with his excellent bird photos. His field notes, banding records, photos, and prints of his book plates are housed at the Western Foundation of Vertebrate Zoology. **Kelly A. Boadway** worked with a variety of seabirds and landbirds as a field technician for 3.5 years in very remote parts of North America, from the Florida Everglades and the French Frigate Shoals of Hawaii to Barrow, Alaska and the High Arctic of Nunavut, where she completed research for her MSc. **Nicole Bouglouan** has watched southwestern France's birds since childhood. After running a business for 20 years, she wrote and illustrated children's fairy tales on bird adventures. Her educational material on the world's birds, along with that of her colleagues, is used by publishers, teachers, towns, naturalist groups, and students. http://oiseaux-birds.com/ **Robert B. Douglas** is Senior Biologist and Forest Science Manager at Mendocino Redwood Company (MRC), which owns and manages 90,000 ha of forestlands in northwestern California. Formed in 1998, MRC's main goal is to manage productive forestlands with a high standard of environmental stewardship while operating a successful business. www.mrc.com **Julie Dufour** was born in the Abitibi-Témiscamingue region of Quebec. A registered nurse in Gatineau,

Quebec, she is passionate about nature photography, especially birds. "I create images so others may appreciate the beauty of the natural world. I try to keep it simple and authentic. In my eyes, simplicity is the ultimate sophistication." **Photography has been **Jenny Erbes's** (jerbes@prbo.org) passion since receiving her first plastic Kodak camera at age seven. She has been working for Point Blue Conservation Science (formerly Point Reyes Bird Observatory) since 2002, monitoring Snowy Plovers and participating in bird, White Shark, and marine mammal research. https://www.flickr.com/photos/48679752@N00/sets/72157629683629287/ **Gail Fraser**, Associate Professor in the Faculty of Environmental Studies at York University in Toronto, studies the ecology, conservation, and management of avian wildlife (e.g., wildlife management and aspects of bird ecology in Toronto's waterfront), and also the management of oil and gas extraction. Her favorite quote: "Support the right to arm bears". http://gsfraser.blog.yorku.ca/ **Born in Edmonton, Alberta, **Robert E. Gehlert** published his first photograph at age 14. Others have appeared in guide books, scientific journals, magazines, newspapers, and calendars. Bob banded birds and studied owls, but worked in fire and safety. Now retired near Tofield, Alberta, he still birds and photographs. **Bosse Haglund:** Biography not available. **Born in Ottawa, Ontario, in 1965, **Karen Hanlon** grew up in various cities in Canada and Germany. She studied zoology at the University of Manitoba and graduated with a BSc. She has worked as a guide dog mobility instructor both in England and Canada, and lives in Perth, Ontario, with her husband and daughter. **Calgary, Alberta wildlife ecologist **Sarah Hechtenthal** specializes in avian biology, conservation, and ecotoxicology. She works as a scientific advisor for aboriginal communities across Canada. She is also a trained oiled-wildlife emergency responder and has participated in several oil spill responses, including the 2010 Deepwater Horizon spill, with International Bird Rescue. www.mses.ca **Born and raised in Aspen, Colorado, **Richard Higgins** studied at the New York Institute of Photography and at Australia's Institute of Photography. Comparing the techniques of several well-known photographers has led him to a new level, and he is metamorphosing from physicist to professional nature photographer. http://www.naturesvisionphotography.com/ **John Hoyt** is a retired PhD computer engineer whose hobbies include fly-fishing, hiking, and photography. He and his wife Donna live in Lewes, Delaware. He regularly volunteers for the Cape Henlopen Hawk Watch near Cape May. John met Robert Alvo on a birding tour of St. Paul Island, Alaska. **Marie-Anne Hudson** is a Bird Surveys Biologist with the Canadian Wildlife Service of Environment Canada in Ottawa, Ontario. She co-founded and then directed the McGill Bird Observatory in Saint-Anne-de-Bellevue for four years. Her PhD at McGill University in Montreal examined avian reproduction on golf courses and urban green spaces. **The late **Edgar T. Jones** started banding birds in 1940 with Ducks Unlimited. By the end of his career, he had banded over 115,000 birds (315 species). A decorated pilot, he operated a bush-flying service after World War II. Later, after unexpectedly acquiring two 16-mm movie cameras, his career as a naturalist and cinematographer began. He produced 10 feature-length films, and his footage appeared on more than 200 nature television programs. In 2001, Edgar was appointed a Member of the Order of Canada for his work in conservation education. **David Laliberte** has been a birder since age 10, and has been photographing birds, plants, and nature for the last 10 years. He has a BSc from Colorado State University, and now lives in St. Petersburg, Florida. His photos have appeared in various books, magazines, and web sites. https://www.facebook.com/david.laliberte.902 **After a friend introduced her to birding in 1976, **Diane Lepage** bought a camera and would rise early to photograph wildlife near Ottawa, Ontario. She later joined the Camera Club of Ottawa. When not working for the federal government, she takes photography trips and also shoots for the Ottawa Field-Naturalists' Club. www.ofnc.ca **A photographer for over 25 years, **Martin Lipman** specializes in corporate, scientific, and natural history work in Ottawa, Ontario. He holds an MA in journalism from Indiana University. Martin is the former academic advisor for the School of Photographic Arts in Ottawa, and continues to work as a private photo educator. http://lipman.photoshelter.com | www.exposuregallery **Craig Machtans**, a Canadian Wildlife Service biologist, has been working with birds professionally for over 20 years. He enjoys photographing birds because it forces patience and often leads to subtle surprises about behavior or interactions—not to mention the simple pleasure of watching the bird rather than hastily completing another survey.
Josef MacLeod and his fiancée Çağdaş Kera Yücel are biology graduate students at Laurentian University in Sudbury, Ontario. Kera found the Common Loon egg (see that account), and Josef photographed it. Kera is an environmental microbial biologist from Fethiye, Turkey; Josef is a freshwater ecologist from Ottawa. They plan to continue working as environmental biologists in northern Canada. **Larry Master** has been photographing nature for about 60 years. After doctoral and post-doctoral studies at the University of Michigan, he worked for 26 years as Chief Zoologist, first with The Nature Conservancy, then with its offshoot NatureServe. He also served on the US Environmental Protection Agency's Science Advisory Board. Larry has numerous publications, including chapters in several books. www.masterimages.org
Alan Murphy's artistic background cultivated his creative use of perches with uncluttered backgrounds and excellent lighting, which became his signature: gorgeous perches paired with beautiful birds showcasing their natural environment with simple elegance. An award-winning photographer with

numerous publications, Alan loves leading workshops. He lives with his family near Houston, Texas. www.alanmurphyphotography.com **Noppadol Paothong** is a wildlife conservation photographer with the Missouri Department of Conservation. His photos have been recognized by more than 60 regional and national awards. He focuses primarily on species at risk. He spent 11 years documenting North American grassland grouse for his book "Save the Last Dance". www.savethelastdancebook.com **Dan Parent** studied psychology at the University of Ottawa in Ottawa, Ontario. An insurance adjuster living across the Ottawa River in Gatineau, Quebec, he works in both provinces. He has always enjoyed nature and wild-life. Photography allows him to express and share his interest in observing wildlife behavior. **Dennis Paulson**, PhD, recently retired from his position as Director of the Slater Museum of Natural History, University of Puget Sound, where he taught biology for 15 years. An expert on both birds and dragonflies, he has led nature tours the world over and has published over 90 scientific papers and 10 books.
Honduran-born Cynthia Pekarik moved to Canada in 1987 to study biology at the University of Guelph. She later released Peregrine Falcons in the Bay of Fundy, studied loons in central Ontario, spent 15 years studying marsh birds and colonial waterbirds on the Great Lakes, and now coordinates international shore-bird and waterbird conservation programs. **Rich Phalin**, a taxidermist in Mukwonago, Wisconsin, loves traveling to different places to photograph wildlife. Special memories include seeing hundreds of Bald Eagles in Homer, Alaska, witnessing loons with newly hatched chicks in northern Michigan, and watching wolves and listening to wolf howls and elk-bugling in Yellowstone National Park. rphalin@wi.rr.com
Frank Phelan completed his MSc (1976) at Queen's University, studying the relationships between hawks and owls and prey populations on Wolfe and Amherst Islands (renowned for owl viewing opportunities in winter). For the last 30 years he has been Manager of Queen's University Biological Station near Chaffey's Locks, Ontario. **Emily Pipher** grew up near Rochester, New York, and completed her BSc at the SUNY College of Science and Forestry. Her MSc thesis at the University of Manitoba examined the effects of grazing intensity on grassland songbird nesting success. She manages a population of Red-cockaded Woodpeckers at central Florida's Archbold Biological Station. **Mark J. Rauzon** is a geography professor, wildlife biologist, photographer, and writer specializing in habitat restoration, especially the removal of feral cats and rats from islands. He is the award-winning author of over 20 non-fiction science books for children, mostly about birds, that are illustrated with his photographs and drawings. www.rauzon.zenfolio.com **Biologist Alain Richard** has long been passionate about photo-graphy. A self-taught birder, he is Past President of le Club d'ornithologie des îles de la Madeleine, Quebec, and often goes on nature outings to appreciate the archipelago's diverse flora and fauna. Many of Alain's photos have been published. alainrichard@yahoo.com | http://www.flickr.com/photos/35495730@N08/
Annie Schmidt has a PhD in ecology from the University of California, Davis. Her research focuses on understanding the unique population responses of different seabird species to changing oceanographic conditions. She is an enthusiastic photographer and filmmaker in her spare time. aeschmidt@gmail.com | http://farallonphoto.blogspot.com **Bill Schmoker**'s photos of over 640 species of North American birds are published in various top magazines, field guides, etc. Bill presents bird photography workshops, gives bird talks, leads field trips, and is a Leica Birding Team member. He is Past President of Colorado Field Ornithologists and serves on the Colorado Bird Records Committee. schmoker.org/BirdPics
François Shaffer, raised in Sherbrooke, Quebec, is a passionate nature observer and a Canadian Wildlife Service biologist in Quebec City. He has been working on the protection of bird species at risk in Quebec such as the Piping Plover, Horned Grebe, Yellow Rail, Chimney Swift, and Peregrine Falcon.
Sam Sheline: Biography not available. **Bryan J. Smith** has degrees in the biological and physical sciences. An avid birder since age six, he has birded and photographed in 47 US states, the Americas, Europe, Africa, and Asia. He lives with his wife Valerie in Tucson, Arizona. His photos are for sale. bryanjsmith@comcast.net | http://www.flickr.com/photos/bryanjsmitheci/sets **Netta Smith** is a research technician in an ophthalmology laboratory at the University of Washington Medical School in Seattle, but her first love is being out in nature, whether skiing, kayaking, hiking, or photographing birds, dragonflies, lizards, and anything else she can find that moves or doesn't. **Lloyd Spitalnik** photographs, bird-watches, leads photo tours, and teaches photography in New York City. His photos appear in Audubon, Natural History, Birder's World, Birding, and Wildbird magazines. He is featured at the Newark Museum. Lloyd helps run the Jamaica Bay Shorebird Festival, runs a rare bird alert, and pursues insects.
Peter Sproule has been an avid birder for 15 years, and has combined his other major interest of nature photography with his interest in all things avian. The advent of digital cameras has made these hobbies even more enjoyable. He shares these passions with his partner Bev. **Arriving in Saskatoon in 1999, Brent Terry** took an immediate interest in birds. After seeing a show featuring Stuart Houston (see bio) studying Great Horned Owls, Brent found one of its nests, contacted Houston, then joined his raptor-banding team. An elementary school building operator, Brent took "his" students on many birding outings before retiring. **Hans Westerlaken** is a high school teacher who conducts breeding bird surveys and waterfowl counts south of Rotterdam. His Dutch list stands at 455 species. Hans travels the world, often in

the company of his wife Caroline, and is hoping to record his 5000th life bird in Peru in 2015. **After retiring from the Tucson Police Department, **Tom Whetten** spent 17 years working for the Arizona Game and Fish Department, where he perfected his photography skills and truly learned about wildlife. He is a professional wildlife/nature photographer and safari guide. www.wildlifephototour.com | https://www.facebook.com/wildlifephototour?ref=hl **Darroch Whitaker** is an ecosystem scientist in Gros Morne and Torngat Mountains National Parks. Born in Newfoundland, he studied at McGill University (BSc), Memorial University (MSc), Virginia Tech (PhD), and Acadia University (Postdoctoral Fellow). He has studied habitat selection, space use, and population biology of birds in managed forest landscapes. **Don Wigle** is a retired epidemiologist living in Ottawa, Ontario. He took up bird photography as a hobby in 2003. http://tinyurl.com/8kzfj2r **Steve Zamek**, a former software engineer, has been birding and enjoying wildlife for several decades. In 2008 he acquired some camera gear, and now endeavors to capture the beauty and behavior of birds in their natural habitat. He lives near productive wetlands along the San Francisco Bay with his wife Jane. www.featherlightphoto.com **Lee Zieger** was Rio Grande Delta Audubon's first President. He has organized four Brownsville International Birding Festivals, and guides birders in the Rio Grande Valley and in Mexico's El Cielo Cloudforest. Photography is his passion, while his community goals are appreciation of the environment and decreased habitat destruction.

REVIEWERS

Stuart Houston, retired radiologist, historian, and Officer of the Order of Canada, has banded, with Mary Houston, 150,000 birds of 211 species with 4000 recoveries of 84 species since 1943. American Ornithologists' Union Vice-President in 1990, he received the Saskatchewan Order of Merit and the AOU Marion Jenkinson Service Award. **Robert L. Manson**—see Cartoonists. **After obtaining his PhD at Yale University, **Raleigh J. Robertson** became Professor at Queen's University in Kingston, Ontario (1971–2010), where he directed its biology station for 33 years and greatly expanded its conservation land holdings. He supervised 47 graduate and 73 undergraduate theses, which contributed considerably to his 200+ papers published on avian behavioral ecology (focusing on mating systems). Raleigh is a Fellow of the Royal Society of Canada. **Jean-Pierre L. Savard** is a Scientist Emeritus with Environment Canada in Quebec City. He obtained his MSc from the University of Toronto and his PhD from the University of British Columbia. His research for the Canadian Wildlife Service (1977–2012) focused on forest birds, urban birds, and sea ducks. jean-pierre.savard@videotron.ca

PROOF READER

Fred Helleiner recently published a book entitled, *For the Birds*, which covers his birding exploits since the 1940s and the concomitant evolution of birding in Canada. He has authored various other birding-related publications. Fred sends out weekly bird sighting summaries for Presqu'île Provincial Park to the Ontbirds/ Birdalert listserv.

DEVELOPER OF MEXICAN BIRD NAMES

Mexican-born **Héctor Gómez de Silva** has lived on three continents and speaks five languages. His PhD thesis examined bird community ecology, but most of his 100+ publications cover Mexican bird distribution, natural history, taxonomy, and conservation. Héctor was involved in raising the Nava's Wren to species status and he translated the voice descriptions in the *Kaufman Field Guide to Birds of North America* to Spanish. He is a bird tour leader, mostly with Eagle-Eye Tours.

POET

Murray Citron is a grandfather and nature-walker who lives and writes in Ottawa, Ontario. His verses have appeared online and in print in Ottawa, Vancouver, Oxford, and New York.

DESIGNER

Suzanne Burkill is a semi-retired digital layout and formatting expert who has designed books covering a wide range of subjects. She continues to keep her hand in by working from home on carefully selected projects. Suzanne lives with her husband Ron near Ontario's birding hot-spot of Presqu'île Provincial Park on Lake Ontario.

DIGITAL IMAGE PROCESSOR

Elizabeth Payne is a freelancing digital color imaging specialist with over 30 years experience working in and teaching her trade. She has three children and three wonderful grandchildren. She loves riding her motorcycle with husband Rick and traveling to warmer places than Ottawa, Ontario, in the winter.

Main Index

*Note: A page number in **bold type** indicates the main entry for that bird.*

Glossy Ibis (*Plegadis falcinellus*), 123
godwits, 192–94
Golden Eagle (*Aquila chrysaetos*), 150, **152**
goldeneyes, 51–52
golden-plovers, 170–71
Gray Hawk (*Buteo plagiatus*), 144
Graylag Goose (*Anser anser*), 15
Great Blue Heron (*Ardea herodias*), 112
Great Cormorant (*Phalacrocorax carbo*), 104
Great Egret (*Ardea alba*), 113
Greater Flamingo (*Phoenicopterus roseus*), 90
Greater Prairie-Chicken (*Tympanuchus cupido*), 75
Greater Sage-Grouse (*Centrocercus urophasianus*),
    66, 67
Greater Scaup (*Aythya marila*), **38**, 39
Greater White-fronted Goose (*Anser albifrons*), 15
Greater Yellowlegs (*Tringa melanoleuca*), 184
Great Horned Owl, 65
Great Salt Lake (UT), 87
Great Snipe (*Gallinago media*), 208
grebes, 83–89
Green Heron (*Butorides virescens*), 119
Greenland, 201
Green Sandpiper (*Tringa ochropusi*), 182
Green-throated Loon (*Gavia arctica viridigularis*), 79
Green-winged Teal (*Anas crecca*), 34
Grey Duck (*Anas s. superciliosa*), 29
Grey Plover. *See* Black-bellied Plover
Griffon Vulture (*Gyps fulvus*), 126
group behaviors (flocking), 58, 97
    cormorants and related species, 100, 101
    herons and related species, 112, 113, 124
    for hunting/foraging, 141, 146, 147
    quails, 60, 63
    raptors (diurnal), 134, 141, 146, 147
    roosting, 113, 134, 146, 183
    shorebirds, 183, 194, 195, 202, 205, 210
grouse, 65–68, 72–74
Grus spp.
    G. americana (Whooping Crane), 164
    G. canadensis (Sandhill Crane), 163
gulls, 55, 96, 106
Gunnison Sage-Grouse (*Centrocercus minimus*), 67
guzzlers, 59
Gymnogyps californianus (California Condor), 128

habitat. *See also* habitat loss; *specific locations*
    agricultural, 14, 159
    alpine, 71, 183, 198
    arctic, 91
    burned, 147, 178
    climate change effects, 38, 154
    coastal, 102, 122, 167, 173, 176, 204, 211–12,
        217
    desert, 60, 62
    of ducks, 14, 26, 56
    estuaries, 43, 54, 103, 116, 197
    fire as management technique, 63
    forests, 58, 68, 134, 137, 140
    fragmentation of, 149

of hawks, 140, 144, 149, 150, 151
    human-created, 102, 121, 124, 149, 204
    of loons, 79
    marshes, 16, 110, 111, 213
    open/grassland, 58, 149–51, 178, 188
    and population, 31
    prairie, 35–36, 191, 194, 213
    protection of, 58, 111
    of quails and related species, 60, 62, 69
    riparian, 140, 144
    sagebrush, 67
    saline, 16, 87, 197, 216, 217
    of shorebirds, 176, 183, 187, 194, 201, 205
    tropical, 149
    tundra/taiga, 17–18, 70, 151, 179, 183–84, 198,
        201
    urban, 138, 141
    wetland, 97, 102, 116, 121, 133
habitat loss
    acid rain as cause, 51, 81
    agriculture as cause, 60, 75
    birds as cause, 17, 102
    coastal areas, 116, 117
    for cranes, 164
    for ducks, 38
    forest-clearing and, 131
    for hawks, 149, 150
    riparian, 140
    for shorebirds, 176, 184, 216
    in wetlands, 97, 111, 124, 133, 154, 159
Haematopus spp.
    H. bachmani (Black Oystercatcher), 168
    H. palliatus (American Oystercatcher), 167
Haliaeetus leucocephalus (Bald Eagle), **135**, 142
Hardy, Fannie, 125
Harlequin Duck (*Histrionicus histrionicus*), 44
Harris's Hawk (*Parabuteo unicinctus*), 141
Hawaii, 190
hawks, 137–51
Heath Hen (*Tympanuchus cupido cupido*), 75
herons, 103, 115–16, 119–21, 214
Himantopus mexicanus (Black-necked Stilt), 165
Histrionicus histrionicus (Harlequin Duck), 44
Hochbaum, Al, 35
Hooded Merganser (*Lophodytes cucullatus*), 53
Hooded Vulture (*Necrosyrtes monachus*), 126
Hook-billed Kite (*Chondrohierax uncinatus*), 130
Horned Grebe (*Podiceps auritus*), 85
House Sparrow, 118
Houston, C. Stuart, 47, 127
Hudsonian Godwit (*Limosa haemastica*), 170, **192**
hummingbirds, 148
hunting (by humans)
    of cormorants, 102, 103
    into extinction, 45, 170
    for feathers, 87, 114
    of grebes, 87
    illegal, 27, 186
    lead poisoning from shot, 37, 128
    legislation controlling, 75

Índice de nombres en español

Index des noms français

About the Author

Left to right: Frances Alvo, Jack Alvo, Robert Alvo, Bob Manson, May 2014

Robert Alvo grew up in Montreal, attended high school in Greece, and studied biology at Queen's University in Kingston, Ontario (BSc). He then obtained his MSc at Trent University in Peterborough, Ontario, where, as Project Biologist of the Ontario Lakes Loon Survey of Long Point Bird Observatory, he studied the effects of lake acidification on Common Loons. This turned out to be a 25-year study, which he published in 2009.

Robert established Canada's first Conservation Data Centre, for Quebec. Later he wrote and translated numerous accounts for the massive provincial bird atlas, *The Breeding Birds of Quebec*. He computerized the Canadian Breeding Bird Census Database and wrote eight national status reports for the Committee on the Status of Endangered Wildlife in Canada (COSEWIC). Robert went on to manage Parks Canada's species database. As a cooperator with the US-based conservation organization NatureServe, he developed Canadian national conservation status ranks for Canada's vertebrates. Finally, the Nature Conservancy of Canada contracted him to negotiate land protection deals in Quebec.

His most important accomplishment in conservation has been to help initiate an international project to develop the Canadian National Vegetation Classification by bringing together Canadian and US experts in the field of ecological classification.

This book marks the beginning of Robert's fifth phase in his evolution of "capturing" species: from young hunter, to observer/photographer, to loon researcher, to conservation status rank developer, to his *Being a Bird in North America* (BABINA) approach.

Robert's favorite hobby is going on birding adventures to new countries. He has been living in Canada's National Capital Region since 1992.

Conserve Canada's Wild Birds!

Photo: Dave Gignac

Join Canada's leading science-based bird conservation organization.

Your gift will support conservation action through critical research and Citizen Science, and programs that engage the next generation in the wonders of birds.

BIRD STUDIES
ÉTUDES D'OISEAUX CANADA

Donate today! 1-888-448-2473 | birdscanada.org